AF877080

Ergebnisse der Plasmaphysik und der Gaselektronik

Band 7

Electric Properties of Weakly Nonideal Plasmas

Ergebnisse der Plasmaphysik und der Gaselektronik

Band 7

Herausgegeben von
Robert Rompe und Max Steenbeck †

Electric Properties of Weakly Nonideal Plasmas

by
Klaus Günther and Rainer Radtke

With 43 Figures and 8 Tables

Akademie-Verlag · Berlin
1984

Authors:

Dr. sc. nat. Klaus Günther
Dr. rer. nat. Rainer Radtke
Akademie der Wissenschaften der DDR, Zentralinstitut für Elektronenphysik, Berlin

Erschienen im Akademie-Verlag, DDR-1086 Berlin, Leipziger Str. 3—4
© Akademie-Verlag Berlin 1984
Lizenznummer: 202 · 100/426/84
Printed in the German Democratic Republic
Gesamtherstellung: VEB Druckerei „Thomas Müntzer", 5820 Bad Langensalza
LSV 1195
Bestellnummer: 763 169 4 (5570/7)

03000

Preface

During the last fifty years the physics of electric arcs culminated in a deep understanding of all essential features of a plasma in local thermodynamic equilibrium on the basis of elementary processes between two particles. It allowed the quantitative calculation of all physical properties of interest as plasma composition, conduction of charge and heat, absorption and emission of radiation in accordance with the experimental experience. About ten years ago the first experimental indications of discrepancies between the calculated and the measured electrical conductivity in thermal plasmas of high charge-carrier density were found. In the following years this observation was established in many laboratories and induced an enhanced investigation of such plasmas which were called 'nonideal' in contrast to the 'ideal' behaviour of plasmas at lower charge-carrier densities. The scientific effort in theory and experiment is as yet increasing, and concerning essential problems we are far from a quantitative understanding.

It is useful for more than one reason to divide the nonideal plasmas into a weakly and a strongly nonideal state. The weakly nonideal plasma is characterized by considerable deviations from the binary collision model due to collective effects which can be taken into account by means of physically justified corrections. In contrast, a strongly nonideal plasma cannot be described in terms of the binary model, and a theory which is able to explain quantitatively such plasmas does not exist until now. Thus, a seperate treatment of the weakly nonideal plasma is justified the more as these plasmas are of great importance in numerous technical applications, e.g. in circuit breakers, welding arcs, pulsed radiation sources, MHD plants etc.

We think that the theoretical and experimental research of the last years on the field of transport processes in weakly nonideal plasmas has yielded results which allow to give a comprehensive and consistent survey. This book explains the physical background of the effects, the experimental techniques, and the results including numerical data of the electrical conductivity in plasmas of different states on the basis of consistent corrections to the binary collision model. We hope that it is useful for the physicist concerned with basic research in the field of nonideal plasmas as well as for the engineer needing quantitative data on the electrical conductivity of high-density plasmas.

Of course, in those areas which are currently developing rapidly, such as dense (nonideal) plasmas, we run the risk of being already slightly out of date at the time of publication. That is most natural; however, we have quoted the very recent papers related to the subjects discussed at the end of this book (section 7.), and we are grateful to the publisher for his extremely pleasant manner of accepting the manuscript in its final version.

It should be noticed that the knowledge and the ideas presented in this work were collected in part during a cooperation with the Institute of Physics Belgrade, Yugoslavia, and we attest gratefully the authorship of S. S. Popović for chapter 5.1. We wish to thank Mrs. M. Dewitz and Mrs. M. Kronisch for technical support in preparing the manuscript and Mr. D. Kubitza † for preparing the computer programs and numerical calculations. Furthermore we are indebted to the editor for his patience in expecting the delayed manuscript and for the quick publication.

K. Günther R. Radtke

Contents

1. Introduction

The kinetic properties of ionized gases are determined by the mutual interaction of their constituents. While in a neutral gas due to the short-range forces of atoms or molecules this interaction is limited to the vicinity of the colliding particles and collisions may be considered to be a sequence of successive events being independent of each other, the presence of charged particles in an ionized gas leads to many new problems. Indeed, due to the long-range COULOMB interactions between charged particles the transport coefficients can no longer be deduced by simple addition of isolated elementary processes which only include the interaction of two particles. However, under certain circumstances the concept of the binary collision theory is applicable also to an ionized gas. In a sufficiently rarified plasma, namely, each charged particle tends to attract a slight excess of opposite-charged particles which lead to a screening of the COULOMB potential at finite distances. This phenomenon may be introduced in the particle kinetics by adapting the DEBYE shielding to dilute plasma conditions where the characteristic length describing this shielding is the DEBYE length D. In this manner, the interaction between the charge carriers may still be taken into account by binary collisions, while depending on the plasma conditions the influence of all other particles is described by a screening space charge. The picture of the DEBYE-shielded COULOMB potential is connected to the assumption of a large number of charged particles in the DEBYE sphere. In high-density low-temperature plasmas, however, this condition is no longer fulfilled and the DEBYE shielding model breaks down. Due to the strong electrostatic interaction between the charged particles, under such conditions one has to take into consideration the whole environment of an ion which confronts us with mathematical difficulties, however. In order to characterize such plasma states the term 'nonideal' (or 'non-DEBYE') is used and, consequently, the possible deviations of real screening mechanisms from the Debye model are called non-ideality effects.

Despite numerous applications — many electrical discharges and plasmas used in science and technology or observed in nature more or less show deviations from the ideal behaviour — these media have been investigated intensively only during the last ten years. This may be explained both by difficulties in precise diagnostics and by problems in the choice of a proper laboratory source for experimental investigation. Thus, with the present state of the art there are more ideas how to describe such media in theory than reliable data concerning transport properites of these plasmas are available. The intention of this review consists in both summing up methods for producing and investigating such plasma states and reporting experimental data about transport properties regarding recent corrections for non-ideality. Thereby, we are restricted mainly to weakly nonideal plasmas in local thermodynamic equilibrium (for the definition of weakly nonideal plasmas see the next chapter). These media represent an important transient condition between ideal and strongly nonideal plasmas, and their investigation is interesting both from physical and technical reasons. To demarcate weakly nonideal plasmas from ideal plasmas on the one hand and strongly nonideal plasmas on the other, in the following some characteristic quantities are introduced by means of which the plasma can be classified with regard to its non-ideality.

1.1. Localization of Weakly Nonideal Plasmas in the Density-Temperature Plane

In the following we shall deal with a partially ionized plasma, the kinetic properties of which are determined by neutral-neutral, charge-neutral and charge-charge interactions. Let U_{pot} be the mean interaction energy of particles and U_{kin} their mean kinetic energy, then a nonideal plasma state occurs if the following condition is violated

$$U_{\text{pot}}/U_{\text{kin}} \ll 1 \,. \tag{1.1}$$

For the range of electron density and temperature considered here, the most important plasma non-ideality is due to the charge-charge interactions whereas non-ideality effects due to neutral-neutral or charge-neutral interactions are possible at much higher particle number densities only. Thus, replacing U_{pot} by $e^2(n_e + n_i)^{1/3}/4\pi\varepsilon_0$ with respect to the COULOMB interaction and U_{kin} by kT the above mentioned ratio reads as follows

$$\gamma = \frac{e^2(n_e + n_i)^{1/3}}{4\pi\varepsilon_0 kT}\,, \tag{1.2}$$

γ being the non-ideality parameter, e the electron charge, n_e the electron particle density, n_i the ion particle density, ε_0 the dielectric constant, k BOLTZMANN's constant, and T the temperature. To simplify matters we assumed to have only one-fold charged ions in the plasma. If manifold charged ions occur we have to take them into consideration in U_{pot}, of course, which can be done easily. Depending on γ we have to distinguish between different degrees of plasma non-ideality. Plasmas, for which $\gamma \ll 1$ are called ideal. Although every interaction between charge carriers more or less results in deviations from the binary collision model, such a classification seems to be justified since the rigorous theory of transport properties which is based on the BOLTZMANN transport equation and the DEBYE shielding indeed can properly be applied to rarefied plasmas. The transient region of weakly nonideal plasmas to be dealt with in the present review is characterized by $\gamma \approx 0.1$, whereas the region of strongly nonideal plasmas is determined by the condition $\gamma \geqq 1$. Besides γ, the number of charged particles in the DEBYE sphere is often used to classify the plasma non-ideality

$$n_D = \frac{4}{3}\pi(n_e + n_i)\, D^3 \tag{1.3}$$

where the DEBYE length D is defined as follows

$$D = \left(\frac{\varepsilon_0 kT}{e^2(n_e + n_i)}\right)^{1/2}. \tag{1.4}$$

Because of $n_D = \gamma^{-3/2}/3\sqrt{4\pi}$ the number of charged particles in the DEBYE sphere decreases if the non-ideality parameter increases. Thus, in case $\gamma \approx 0.1$ (weakly nonideal plasma) we obtain $n_D \approx 3$, and the approach of the DEBYE-shielded COULOMB potential which is connected with the assumption $n_D \gg 1$ is inadmissible. On the other hand it is evident that for $n_D \gg 1$ (these plasmas are sometimes called Debye plasmas) the condition $\gamma \ll 1$ holds and the plasma is ideal according to our classification. Figure 1.1 indicates the range of electron density and temperature where nonideal plasmas exist. Of course, in this large area besides influences due to nonideal charged particle interactions many other effects have to be considered which lead to very complex phenomena.

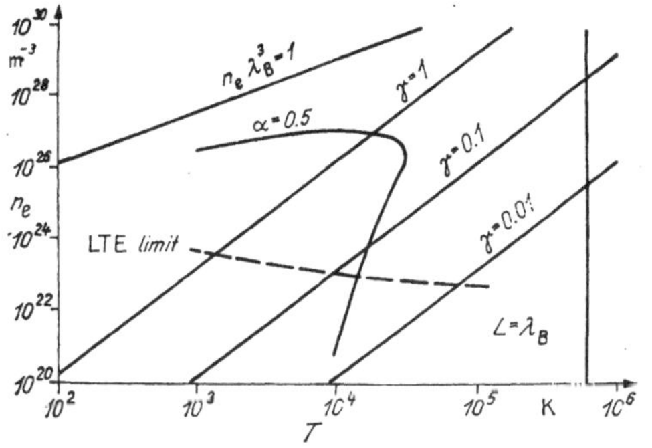

Fig. 1.1. Localization of plasma parameters in the density-temperature plane

For long range potentials like the COULOMB, at high energies (or temperatures) the quantum character of the charge carrier interaction becomes important and the classical treatment of the corresponding scattering problem is insufficient. If for the effective interaction length of charged particles the LANDAU length $L = e^2/4\pi\varepsilon_0 kT$ is used, then the temperature around which the transition between classical and quantum character of particle interaction occurs may be estimated from $L \approx \lambda_B$ where $\lambda_B = h/(2\pi m_e kT)^{1/2}$ is the thermal DE BROGLIE wavelength. As shown in figure 1.1, this transition is independent of the electron density and will appear only at comparatively high temperatures. We should mention also the region of degeneration $n_e\lambda_B^3 > 1$ where for the electrons the BOLTZMANN statistics has to be replaced by the FERMI statistics. However, as can be seen from the figure, this transition occurs above $\gamma \approx 1$ outside the parameter range of present interest.

Contrary to quantum effects which influence the properties of weakly nonideal plasmas at higher temperatures only, the demand for local thermodynamic equilibrium (LTE) leads to a drastic restriction of the temperature and density range being of interest here. To demonstrate this, in figure 1.1 the minimum electron density required for LTE conditions is plotted where validity criteria according to GRIEM [Gri63] and DRAWIN [Dr69] were used regarding recent recommendations given in [CHR75]. Since the curve dividing the $n_e - T$-plane in figure 1.1 into the LTE and non-LTE region was calculated for hydrogen, we have to expect a shift of this limit in going from hydrogenic to non-hydrogenic systems. Thus, the range of validity of the LTE model as indicated in the figure cannot be transferred to other plasmas but it illustrates the necessity to consider questions of equilibrium conditions if the electron density falls below 10^{23} m^{-3}. Particularly in stationary arc discharges, a possible departure from LTE should be taken into consideration at least for the outer parts of the discharge to avoid serious errors in the interpretation of experimental results.

The complexity of various elementary processes in weakly nonideal plasmas becomes apparent considering the ionization equilibrium in the remaining part of the $n_e - T$ plane. If α denotes the degree of ionization, then the general behaviour of the plasma may qualitatively be discussed by introducing the curve $\alpha = 0.5$ into the density-temperature diagram. In figure 1.1 this was done for hydrogen using results presented in [EKK76]. Outside $\alpha = 0.5$ there is a nearly fully ionized gas whose kinetic properties are essentially determined by charged particle interactions ($e - i$ and $e - e$ collisions). With decreasing α, the influence of charged particle interactions is still dominant due to the long-range COULOMB potential, but depending on the magnitude of the corresponding cross-section the interactions between electrons and neutrals ($e - a$ collisions) may lead to significant quantitative corrections of the macroscopic properties. Besides problems arising from the breakdown of the DEBYE shielding theory, the correct inclusion of $e - a$ collisions, therefore, is one of the most important questions concerning the theoretical description of high-density low-temperature plasmas. At high densities we notice a turning down of the curve $\alpha = 0.5$ to the left which is connected with the growing influence of charged particle interactions. According to [EKK76], where a theory of the ionization equilibrium was developed from a very general point of view, this region is characterized by a rapid change of α and may be considered to be a diffuse transition from a partially ionized plasma to a completely ionized metallic state. Under certain circumstances a phase transition to a highly conducting state also occurs if at high non-ideality the condition for the thermodynamic stability is violated leading to first order phase transition predicted by NORMAN and STAROSTIN [NS68]. For a detailed survey concerning typical non-ideality effects at $\gamma \geqq 1$, the reader is referred to [EKK76].

1.2. Occurrence of Weakly Nonideal Plasmas and Phenomena Expressing Non-Ideality

As shown in the last chapter, nonideal plasma states are classified by high electron particle densities and low temperatures. Considering the corresponding area in figure 1.2, nonideal conditions are found in many electrical discharges and plasmas used in science or technology. So, the region $\gamma \geqq 0.1$ includes plasmas produced by stationary or pulsed high pressure arcs [Bau76, BDM73, Beh71, BM71, BR70b, GLR83, HMU72, Ko71a, PS70, RG76], shock waves [BLSF75, FLDG76], ballistic compressors [DDES69 DH78, KK74a], pulsed capillary discharges [EK74, OPS77], wire explosions [DZ77], etc. In many devices the non-ideality is due to the specific properties of such plasmas and special attention is directed to possible applications in the plasma technology rather than to the systematic investigation of non-ideality effects. Examples are high-intensity radiation sources, circuit breakers, and a combination of a nonideal plásma MHD generator with a fission reactor [BKN77]. Concerning basic research in the field of nonideal plasmas, at present there exist two main types of approach: firstly, the creation of a laboratory source with strong nonideal conditions ($\gamma \approx 1$) but very

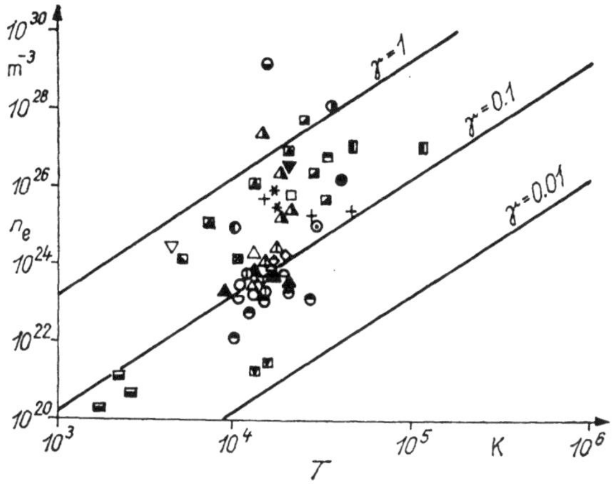

Fig. 1.2. Values of electron density and temperature;
explosive shock compression: ◒ — Cu [KKPR77];
stationary high-pressure arc: ▲ — Ar [Bau76, BFDM73], ⌓ — Ar [Ko71a], ○ — Ar [GNPP78], ⦶ — Ar [BM71], ◓ — H [PS70, Beh71], ◡ — Xe [HMU72];
pulsed high-pressure arc: △ — Ar [GLR83], ⟁ — Ar [BR70b], + — Ar [MKG75], ✳ — He [MKG75], ◇ — H [RG76];
pulsed capillary discharge: ⊙ — plexiglas [EK74], ● — plexiglas [OPS77], ▲ — Cs [KMRT73];
shock waves: □ — Cs [BLSF75], ◪ — Xe [FLDG76], ▲ — Ar [FLDG76], ▲ — Ne [FLDG76] ⊠ — air [FLDG76];
wire explosion in high-pressure atmosphere: ▬ — Cs [DZ77];
adiabatic compression: ▽ — Cs [KK74a], ▼ — noble gases [DH78], ◐ — Xe [DDES69];
pulsed-discharge in liquids: ◑ — H_2O [Rob67], ◩ saturated NaCl-water solution [DRRZ77];
decaying plasma (afterglow): ▼ — Ar [SH72b];
oven-generated plasma: ▬ — Cs [EKKR71];
exploding gaseous filament: ◧ — air [SRG77];
optical discharge: ⧨ — Ar [GZKM70, GZKM71]

poor precision in diagnostics [AK77] and, secondly, the development of diagnostic and data acquisition methods parallel to the development of the plasma source without special emphasis on the extent of non-ideality. Thus, making strict demands for the homogeneity and time development of the plasma under investigation only a few reliable and completely controlled laboratory devices are available for precision measurements in nonideal plasmas. As can be seen from the examples in figure 1.2, plasmas generated by both stationary and pulsed wall stabilized high-pressure arcs are predominantly situated near $\gamma \approx 0.1$. This is the region of weakly nonideal plasmas. The investigation of such plasmas is necessary both for physical and technical reasons: they represent an important transient condition between the ideal and the strongly nonideal plasmas, and their technical applicability depends on the precise knowledge of their properties. Of course, the quantitative investigation of weakly nonideal plasmas necessitates precise measurements in well-defined and reproducible plasmas using diagnostic methods whose interpretation is not or in a calculable manner influenced by non-ideality effects. Furthermore, the calculation of transport coefficients from macroscopic properties requires the reduction of integral experimental data to specific data related to defined thermodynamic conditions of the plasma. Obviously, the reliability of such calculations increases if the spatial symmetry of the plasma is as high as possible and can be measured directly in the experimental arrangement. The possibilities to produce weakly nonideal plasmas shall be revised in chapter 3. with special attention to the fulfilment of these requirements.

As demonstrated in figure 1.3 on the basis of equilibrium composition calculations in the case of argon, weakly nonideal conditions occur in plasmas at almost any

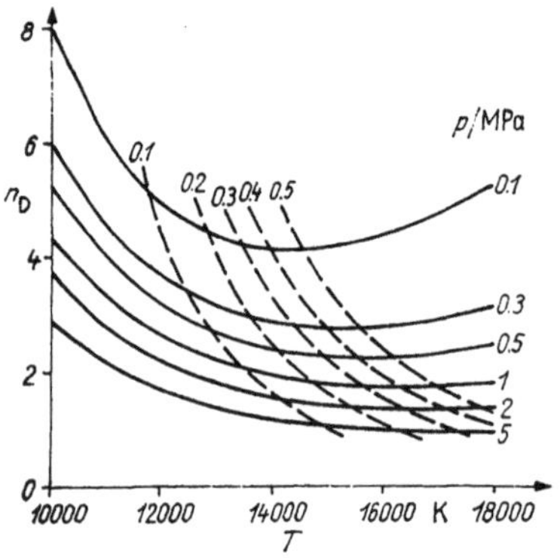

Fig. 1.3. Temperature- and pressure-dependence of n_D in argon plasmas (— — — curves of equal degree of ionization)

degree of ionization. In the investigation of non-ideality effects, however, it is advis able to prefer plasmas with a degree of ionization as high as possible for the following reasons: first, the rate of convergence of approximations in theoretical calculations of transport coefficients depends on the plasma composition and falls with decreasing α due to the introduction of $e - a$ collisions, and secondly, the interpretation of experimental data is rendered more difficult by uncertainties connected with the existence of atoms (choice of electron-atom cross-sections including excited states) since at low degrees of ionization non-ideality effects which are connected with the electron-ion interaction are completely hidden by $e - a$ collisions. Bearing in mind these facts, the high electron density necessary for nonideal conditions should be established at slightly raised temperature but lower neutral particle densities rather than at elevated pressures as was done for example in the plasmas investigated by BAUDER [Bau76, BDM73].

Nonideal plasmas are of interest not only from the basic research or plasma technology aspect but also in astrophysical investigations. So, the atmosphere of hydrogen-rich dwarf stars ($T \approx 10^4$ K, $n_e \approx 10^{23}$ m^{-3}) and the base of the solar photosphere are examples for weakly and strongly nonideal plasmas occuring in nature.

Considering the large variety of thermodynamic, transport, and radiative properties affected by charged particle interactions, non-ideality effects should become apparent by a series of phenomena resulting in significant corrections compared with theoretical plasma models based on the Debye shielding theory. However, in partially ionized laboratory plasmas effects of non-ideality appear together with a number of other effects already mentioned or several of them compensate each other. Consequently, a careful analysis of overall experimental data is necessary selecting nonideality effects from a number of other phenomena.

The first physical effect due to non-ideality reported in the literature was the discrepancy between the electrical conductivity measured in high power discharges (see, e.g., [Bau76, BM71, BR70a, BR70b, EKKR71, GNPP78, GPPR76, GR72, PPV74, Rov72a, VP72]) and the numerical results obtained from a CHAPMAN-ENSKOG solution to the BOLTZMANN transport equation (see, e.g., [CSR50, SF58a, SF58b, SH52, SJB66, Sp56]). As figure 1.4 illustrates, this difference is the more pronounced the

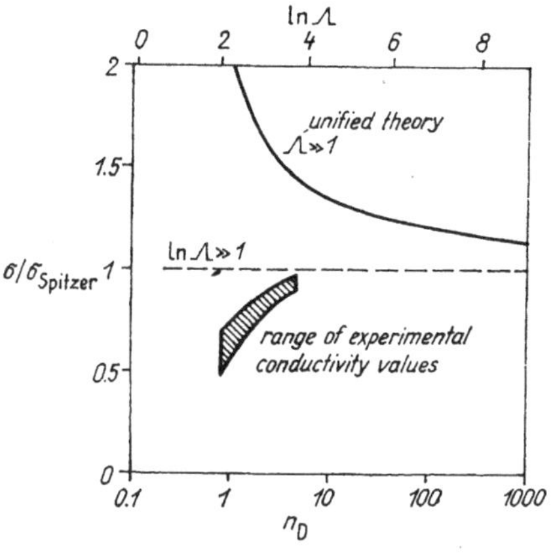

Fig. 1.4. COULOMB part of the electrical conductivity as a function of DEBYE number n_D and ln Λ related to SPITZER's conductivity. Dashed line: extrapolation of the SPITZER calculation to low ln Λ-values

smaller the number of charged particles in the Debye sphere is. The straight line $\sigma/\sigma_{\text{Spitzer}} = 1$ represents an extrapolation of SPITZER's conductivity formula [Sp56] which holds for fully ionized plasmas characterized by $\ln \Lambda \gg 1$.

$$\Lambda = \frac{12\pi\varepsilon_0 DkT}{e^2} \tag{1.5}$$

is the ratio of the DEBYE length and the mean impact parameter for a 90° deflection in an $e - i$ collision. Another dimensionless parameter which is characteristic of the plasma non-ideality is the so-called plasma parameter

$$\mu = \frac{e^2}{4\pi\varepsilon_0 DkT}. \tag{1.6}$$

Between the different parameters of non-ideality, the following relations hold

$$\Lambda = 9n_D = 3/\mu \; . \tag{1.7}$$

Within the framework of the 'unified theory' which is connected with the less restrictive condition $\Lambda \gg 1$ (see, e.g., [GW67, It63, KA63, KAI63, Vi71, WW69]), some corrections were obtained allowing to extend the range of validity of the classical theory of SPITZER to small values of the COULOMB logarithm. As indicated in the figure, these so-called 'small ln Λ' corrections result in higher conductivities tending to confirm the measurements of BAUDER [Bau76] not presented in this drawing. However, in the region of small n_D this approach becomes divergent and any comparison with experiments may lead to serious errors in the interpretation. To correct the shortcomings of the original WILLIAMS-DEWITT treatment [WW69], in a very recent paper ROGERS et al. [RWB81] solved the kinetic equation of GOULD and DEWITT [GW67] including plasma screening effects as well as scatterings from neutral particles. As a result, the electrical conductivity falls well below the predictions of 'unified theory', and reasonable agreement with measurements in strongly nonideal plasmas is obtained.

Many attempts have been made to understand and explain the nonideal plasma transport properties. The different approaches developed to this problem (see section 2.3.) lead to corrections for the non-ideality which more or less agree with experimental results. However, regarding the approximative character of these calculations, it is difficult to give a comprehensive explanation to all phenomena observed in weakly nonideal plasmas.

Considering similarity laws for Coulomb components of thermodynamic, transport and radiative plasma properties, Kulik et al. [KKN71, KKR72, Ku72] have shown the possibility of further effects caused by the incomplete shielding of charged particles at the Debye length. Of course, such a simple treatment does not take into consideration all peculiarities of nonideal plasmas but it allows to form a rough idea concerning the deviations from the ideal behaviour to be expected in the experiment. For example, an estimate of electrostatic corrections to the different thermodynamic functions suggests only a minor influence of charge carrier interactions on relations which describe the thermodynamic state of the plasma. Thus, due to the smallness and a partial compensation of nonideal corrections to pressure and electron density the investigation of the ionization equilibrium seems to be of little benefit in studying non-ideality effects in weakly nonideal plasmas.

The radiative properties of nonideal plasmas have attracted much attention within the last few years, since they are expected to be influenced significantly by effects related to the high values of electron concentration. A topic of some controversy, both experimental and theoretical, is the transparency window predicted in a series of papers (see, e.g., Kobzev et al. [KKN77], Kobzev and Kurilenkov [KK78] and Vechinin and Kurilenkov [VK80]). This effect relates to the oscillator strength distribution around the series limits which should exhibit a gap owing to the high electron concentration under nonideal conditions. First experiments dealing with this phenomenon were carried out by Kurilenkov and Minaev [KM78] with mercury and gave an apparent verification of the proposed model. However, reference to the more reliable theoretical treatment of this problem (Höhne and Zimmermann [HZ82]) indicates that a discontinuity does not appear at threshold which is in full agreement with recent experimental studies on high-pressure hydrogen arc discharges (Radtke and Günther [RG]). Concerning the high-pressure peculiarities of continuous radiation much theoretical and experimental work was done during the recent years (see, e.g., [BM71, GC77, GCD75, GNP77, KK74b, KK78, KKN77]). Besides finding well-known high-density plasma effects (e.g. advance of the apparent series limit), several authors observed an unexpected variation of the intensity with pressure. So, for the near infrared continuum emission of high pressure ionized argon, Batenin and Minaev [BM71] found values being in excess of ideal plasma calculations. In consideration of the wide congruency observed in the behavior of radiative and transport processes the authors assumed non-ideality effects to be responsible. However, considering recent experimental studies of IR continuum radiation [GC77] which confirm ideal plasma models even in the case of $n_D \approx 2$, it is not clear at present whether the high pressure peculiarities can be attributed to nonideal plasma effects or to possible insufficiencies in the diagnostics of dense plasmas. Concerning line shape observations in the region $\gamma \geqq 0.1$, no systematic studies have been made although the nonideal micro-field distribution substantially deviates from ideal plasma calculations (see, e.g., [KF76b]) and modifications to the overall line shape are most likely. Recent improvements of usual Stark broadening calculations [Gri78, Ros75, Se77a, Se77b, WKH75] remove an experiment-theory discrepancy as found especially for Lyman and Balmer hydrogen lines (see, e.g., [GW77, GW78, WKP72]) at relatively low degrees of nonideality ($n_D \approx 5$). However, the poor agreement between theory and experiment noticed by Baker and Burgess [BB79] for a number of line shape features, particularly at the higher values of n_e, has shown some evidence of high-density effects which are not yet identified.

Turning once more to plasma transport properties, it is noticed that effects of nonideal charged particle interactions were observed only for the electrical conductivity but not for other coefficients as thermal conductivity or viscosity. With regard to the latter, this is due to considerable difficulties in the evaluation of this quantity even at less restrictive low-pressure conditions (concerning the details the reader is referred to [Dr68]). As a consequence, only a few measurements of the plasma viscosity have been reported in the literature [AC70, AP70, CLL67, SHB71] for pressure ranges being of minor interest in studying dense plasma peculiarities.

Concerning the thermal conductivity in plasmas at elevated pressures, the situation is rather complex due to the transport of energy via radiation. Indeed, most of the measured conductivity values are the sum of two contributions indistinguishable from each other by experiment: the ordinary thermal conductivity due to both diffusion and reaction processes and the effective radiative thermal conductivity due to radiative diffusion. Only the former is unambiguously related to the thermodynamic state of the plasma whereas the latter depends on the plasma state variables as well as the spatial dimensions of the plasma. At pressure levels considered here it must be expected that the radiative thermal conductivity gives non-negligible contributions so that a laborious procedure is necessary to extract the electron component of thermal conductivity from measured plasma characteristics (see, e.g., [Ko71b]). Moreover, since the radiative thermal conductivity itself is affected by non-ideality effects as may be concluded from recent investigations of high pressure Cs-plasmas [KRR77], it is probably difficult, if not impossible, to draw reliable conclusions on nonideal charged particle interactions on the basis of thermal conductivity measurements. Bearing in mind these facts, much preference is given to the electrical conductivity which is easier to handle due to its direct relation to current and field strength. However, because of the complicated connection between the various elementary processes and their macroscopic result — the observed electrical conductivity — further information concerning the standard approaches used in deriving electron transport properties from overall experimental data is necessary.

2. Description of Ideal and Weakly Nonideal Plasmas

One of the most important problems in the treatment of ionized gases is to adequately describe the interaction between charged particles. In particular, the long-range nature of the Coulomb potential necessitates consideration not only of both colliding particles but also of their environment. In low-density (ideal) plasmas, screening effects can be taken into account in the calculation of transport properties by adapting the Debye shielding theory to plasma conditions. However, at high values of electron concentration this approach breaks down and new problems arise: the establishment of the shielding equilibrium takes nearly as long as the mean time between two collisions, and the strong electrostatic interaction of particles probably leads to new compounds (localized electrons, collisional complexes) which may influence the macroscopic properties considerably.

To seperate eventual non-ideality effects from overall experimental data, a detailed and careful analysis of all influences which could produce the same quantitative consequences as the violation of ideal plasma assumptions is necessary. This includes the consideration of peculiarities in the production and diagnostics of weakly nonideal plasmas which will be dealt with in chapters 3. and 4. as well as a careful analysis of the theoretical models used in deriving electron transport coefficients. Concerning the latter aspect, many basic calculations are carried out assuming ideal plasma conditions and non-ideality effects are introduced through corrections of the classical binary collision theory. Hence, before dealing with (weakly) nonideal plasma transport properties, in section 2.2. we are concerned with the theoretical background in treatments of ideal plasmas.

In section 2.3. special attention is paid to nonideality effects both from the experimental and theoretical point of view. Although many attempts have been made to give a theoretical explanation of transport phenomena observed in high-density low-temperature plasmas, a rigorous description of such media cannot be considered as completely established. So, the different nonideality corrections to be dealt with in section 2.3. will certainly have some influence on the transport properties but a kinetic theory that coincidentally involves all these effects is missing up to now. In that way the present work reflects the momentary state of the knowledge and it is up to further investigations to make a choice between the different models proposed or to suggest an improved one.

Inasmuch as the transport coefficients depend sensitively on the thermodynamic plasma state variables, in the following some information is given concerning this aspect of the computations and the magnitude of high density corrections to be expected in equilibrium composition calculations is shown.

2.1. Equilibrium Plasma Composition

The quantitative investigation of transport properties necessitates information on the plasma composition as a function of temperature and pressure. Since in weakly non-ideal plasmas local thermodynamic equilibrium prevails, the composition may be computed in the usual way from the mass action law for ionization and the equation of state taking into account high-density corrections to both the ionization energy and the kinetic pressure. According to thermodynamic considerations, the ionization potential decrease and the pressure changes result from the corresponding high-density corrections to the free energy and read in the case of DEBYE approximation as follows (see, e.g., [EKK76])

$$\Delta E_i = \frac{e^2}{4\pi\varepsilon_0 D}, \tag{2.1}$$

$$\Delta p = -\frac{kT}{24\pi D^3}. \tag{2.2}$$

More refined treatments [Günd70, Günd71, EK66a, EK66b] accounting for quasi-free highly correlated particles resulted in slightly different expressions for the lowering of the ionization potential. So, ECKER and KRÖLL [EK66b] find

$$\Delta E_i = 1.4 \frac{e^2}{4\pi\varepsilon_0 D} \tag{2.3}$$

which is in numerical agreement with the formula derived from GÜNDEL's concept [Günd70, Günd71] if we restrict the electron density to the range below 10^{24} m^{-3}. Whereas in ideal plasmas the dominant corrections to the free energy are due to plasma screening effects, in nonideal ones it must be corrected additionally on account of short range interactions (charge-charge, charge-neutral and neutral-neutral collisions). Therefore, the above expressions represent only the first term of a series expansion for deviations from the ideal gas behaviour to be expected due to particle interactions. To estimate the magnitude of further contributions various methods have been developed in recent years (see, e.g., [Di79, EKK76, ZKN76, ZNF75a, ZNF75b]). In [EKK76] the thermodynamic functions were determined quantum-statistically assuming that non-ideality is caused by the effect of COULOMB forces only. A useful approximation describing the high-density corrections to the ionization energy and the pressure exactly up to order $O(n_e^2)$ reads

$$\Delta E_i = \frac{e^2}{4\pi\varepsilon_0 D} \cdot \frac{1}{1 + \sqrt{\pi}\lambda_B/4D}, \tag{2.4}$$

$$\Delta p = -\frac{kT}{24\pi D^3}\left(1 - \frac{3\sqrt{\pi}\lambda_B}{8D} + \frac{9\pi\lambda_B^2}{80D^2} - \frac{\pi^{3/2}\lambda_B^3}{32D^3}\right) \tag{2.5}$$

where λ_B is the thermal DE BROGLIE wavelength. Independent work on the evaluation of high density corrections was done by ZELENER and co-workers [ZKN76, ZNF75a, ZNF75b] who investigated the equation of state in the region $n_D = 0, 1 \ldots 1$ taking as a basis thermodynamic pseudopotentials for the charged particle interactions. Their numerical results may be represented by an expression that approaches the relation (2.1) in the limit of small densities:

$$\Delta E_{\mathrm{i}} = \frac{e^2}{4\pi\varepsilon_0 D}\left(1 - \frac{1.052}{4\pi n_{\mathrm{D}}} - \frac{0.74}{48\pi^2 n_{\mathrm{D}}^3} + \frac{0.42}{576\pi^3 n_{\mathrm{D}}^5}\right). \tag{2.6}$$

Recent calculations [Di79] finally have extended the range of validity of the preceding approximations including contributions due to atom-atom interactions.

To show the influence of high-density corrections on equilibrium composition calculations, in figure 2.1 the ratio $(p_{\mathrm{id}} - p)/p_{\mathrm{id}}$ is plotted as a function of n_{D}. p corresponds to the pressure calculated from the mass action law for ionization and the

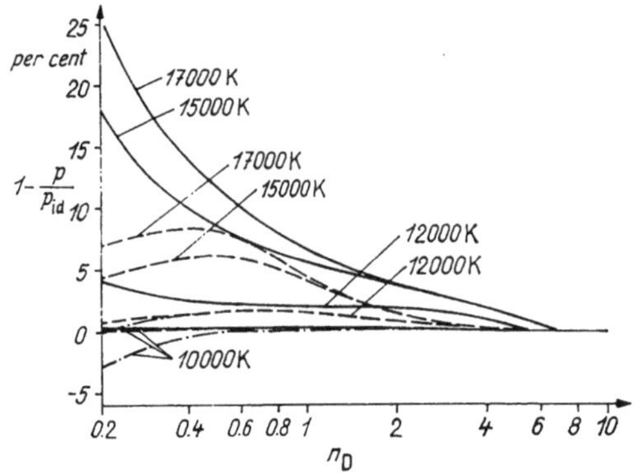

Fig. 2.1. Influence of high-density corrections on equilibrium composition calculations for xenon according to Ecker and Kröll [EK66b] (———), Zelener et al. [ZNF75b] (— — —), and Dienemann [Di79] (—·—·—)

equation of state taking into account the respective high-density corrections, whereas p_{id} represents the ideal gas pressure computed supposing $\Delta E_{\mathrm{i}} = 0$ and $\Delta p = 0$. The results of Ebeling et al. [EKK76] have been suppressed in this comparison since no essential differences with another calculation (Zelener et al. [ZKN76, ZNF75a, ZNF75b]) could be found on the presented scale. As can be seen from the figure, apart from deviations caused by neglecting the quasi-free particles in the frame of models [Di79, EKK76, ZKN76, ZNF75a, ZNF75b] all methods we dealt with here are equivalent in the range $n_{\mathrm{D}} > 0.5$. In the region of stronger non-ideality ($n_{\mathrm{D}} < 0.5$) two phenomena are noteworthy: first, owing to a partial compensation of non-ideality corrections to pressure and electron density, there is only a comparatively minor influence of charge carrier interactions on equilibrium composition calculations, and secondly, with decreasing temperature possible deviations from the ideal gas behaviour are due to neutral-neutral interactions rather than to charged particle interactions.

As can be concluded from the investigations quoted before, the demarcation of free and bound states becomes the more difficult the larger the perturbation by the plasma is. Ecker and Kröll [EK66a] demonstrated that a discontinuous limit between them does not exist and discrete energy states with small binding energies may be considered to be free as well as bound. Consequently, in this region we cannot distinguish between different particle kinds which has implications not only on equilibrium composition calculations but also on transport properties. In a series of papers Vorobev and Khomkin [Kh74, Kh78, VK76, VK77a, VK77b] used the concept of quasi-free particles to calculate the kinetic properties of weakly nonideal plasmas. As will be shown below one reason for the non-ideality observed in the behaviour of the electrical conductivity could be the so-called collision complexes which represent correlated electron-ion pairs of positive interaction energy.

2.2. Transport Processes in Ideal Plasmas

The presence of forces in plasma which may be due to gradients in one or more of the parameters (tmeperature, densities etc.) or caused by external fields gives rise to various transport phenomena. If the forces are assumed to be small, there is a linear connection between them and the flux vector associated with the property transported. The proportionality factors between the flux vector and the various forces are the transport coefficients. For instance, if an electric field $\boldsymbol{E}$ and a temperature gradient ∇T is present in plasma, there is an electric current with the density

$$\boldsymbol{j} = \sigma \boldsymbol{E} + \varkappa \nabla T . \tag{2.7}$$

In our example, σ is the electrical conductivity and $\varkappa$ the thermal electric coefficient. Further transport coefficients as the thermal conductivity or viscosity are due to the transport of kinetic energy and momentum, respectively.

Various conceptions have been developed in deriving transport properties from the elementary processes taking place in plasma. In the elementary theory formulas were obtained that relate the transport coefficients to a linear combination of certain weighted transport cross-sections. So, neglecting the contribution of ions, a simple expression for the electrical conductivity reads

$$\sigma = \frac{e^2 n_e}{m_e \langle \nu_e \rangle} \tag{2.8}$$

where

$$\langle \nu_e \rangle = \sum_j n_j \langle Q_j v_e \rangle \tag{2.9}$$

is the effective collision frequency due to collisions of electrons with particles of type j and $Q_j(v)$ are the corresponding velocity dependent transport cross-sections. The brackets indicate that an ensemble average has to be taken in computing $\langle \nu_e \rangle$. Strictly speaking, formula (2.8) applies for a Lorentz gas but often a properly chosen transport cross-section Q_e was added in the denominator to account also for electron-electron interactions. However, since these encounters have an effect only in the presence of heavy particles — otherwise the total change of momentum is zero in such collisions — they cannot simply be described by extending the formalism used for electron heavy-particle collisions. In a series of older papers (see, e.g., [BPR67, Dr56, FM56, MPS52]) the elementary theory has been employed to derive electron transport properties from electric arc measurements. However, due to the conceptual problems involved in this model it is difficult to draw reliable conclusions about particle interactions in ideal plasmas not to mention in nonideal ones.

In kinetic theory of transport processes, the macroscopic properties are described by the distribution function f_j which contains information on both the spatial distribution of particles and the distribution with respect to their velocities. Normally, it can be divided into an isotropic part with the equilibrium Maxwellian distribution and a directional part expressing the dependence of f_j on the external sources and/or the gradients in the plasma parameters. The standard approach to determine the distribution function is based on solving the Boltzmann transport equation using the Chapman-Enskog formalism [CC52]. Since it is well introduced in plasma physics, in section 2.2.1. we shall give only a brief account of it.

Given the distribution function, it is simple to describe the various transport processes by it. So, considering the passage of particles j through the plasma each of them carrying the property P, the corresponding flux vector can be written [FK72]

$$\boldsymbol{j}_P = \sum_j \int P f_j(\boldsymbol{v}, \boldsymbol{r}, t)\, \boldsymbol{v}_j\, d\boldsymbol{v}_j \,. \tag{2.10}$$

If we succeed in expressing (2.10) in terms of the forces present in plasma, the transport coefficients are given by the corresponding proportionality factors. The flux vector associated with the transport of electric charge can be obtained from (2.10) by setting $P = e_j$ where e_j denotes the electric charge of the species j. In determining the electric transport coefficients it is usual to account only for the contribution of electrons because the ions may be considered to be at rest owing to their relatively large mass.

As regards the thermal conductivity, the situation is frequently confused due to the underlying theoretical complexity. Indeed, setting $P = m_j/2\,v_j^2$, the lefthand side of (2.10) can be interpreted as the heat flow, but obviously it does not contain contributions from both the reactive and the radiative thermal conductivity. The calculation of the latter necessitates consideration of plasma radiative properties which, however, is beyond the scope of the present work. The reactive thermal conductivity is due to ionization and/or dissociation reactions and can be expressed as linear combination of certain binary collision integrals.

Some other approaches to the problem of the electrical conductivity have been developed recently within the frame of a quantum-statistical many body theory ([BKM80, ER79, MR82, REK80, RMKK79]). Röpke et al. [RMKK79] and Blümlein et al. [BKM80] actually derived expressions for the complex dielectric function $\varepsilon(\omega)$ which is related to the conductivity by

$$\sigma(\omega) = \frac{\omega}{4\pi i}\, \{\varepsilon(\omega) - 1\} \,. \tag{2.11}$$

Within this concept approximations were obtained for the reduced electrical conductivity containing the general dependence features of the dielectric function.[1]) The numerical results show Spitzer-like behaviour in the limit $\omega \to 0$, but a spreading-out occurs for the σ^*-curves at different temperatures even if the plasma is ideal according to our classification. One plausible reason for this difference might be associated with the arbitrarily chosen screening length within the kinetic approaches.

An alternative calculation was done by Ebeling and Röpke [ER79] and Röpke et al. [REK80] who investigated the plasma conductivity applying recent concepts and results of kinetic theory (see, e.g., [Kli75]) as well as the Kubo-type correlation function theory (see, e.g., [Sub76]). The calculations account especially for effects of dynamical screening, short-range forces and the relaxation of the ion cloud in the

[1] The dimensionless reduced conductivity is defined by

$$\sigma^* = \frac{e^2 m_e^{1/2}}{(4\pi\varepsilon_0)^2\,(kT)^{3/2}}\,\sigma\,,$$

and the related expression for the conductivity of a fully ionized plasma according to Spitzer [SH52] may be written in the form

$$\sigma^*_{\mathrm{Sp}} = 2(2/\pi)^{3/2}\, 0.5816\, (\ln \Lambda)^{-1}$$

predicting a dependence of σ^* only on a single parameter Λ (or μ).

presence of an external electric field (DEBYE-ONSAGER relaxation effect). The related corrections have only a small effect on the conductivity of weakly nonideal plasmas but they greatly improve the agreement with experiments in strongly ones.

Another method to evaluate electron transport coefficients are the molecular-dynamical calculations of NORMAN and VALUEV [NV79, VN77]. This approach is based on the direct computation of electron trajectories and may be considered as computer experiment to test any theoretical model developed to the problem of plasma transport properties.

2.2.1. The Transport Equation

Any kinetic theory dealing with the calculation of transport coefficients rests upon the BOLTZMANN transport equation which in its most general form may be written

$$\frac{\partial f_l}{\partial t} + \boldsymbol{v} \operatorname{grad}_{\boldsymbol{r}} f_l + \boldsymbol{K} \operatorname{grad}_{\boldsymbol{v}} f_l = \sum_j \left(\frac{\delta f_l}{\delta t}\right)_j \tag{2.12}$$

where $f_l = f_l(\boldsymbol{v}, \boldsymbol{r}, t)$ is the distribution function for particles of type l interacting with particles of different types j, $\boldsymbol{K}$ is the external force per unit mass, $(\delta f_l/\delta t)_j$ the change in f_l due to collisions with particles of type j, and the remaining symbols have their usual meaning. The streaming part (l.h.s. of (2.12)) describes the motion of particles along their trajectories in phase space, while the collision term is intimately connected with the mechanism of particle interactions.

In slightly ionized gases, the dominant interaction process is due to collisions between electrons and atoms, and the BOLTZMANN form of the collision term can be introduced into (2.12). It may be written as follows [FK72]

$$\left(\frac{\delta f_l}{\delta t}\right)_j^{\mathrm{B}} = n_j \int_0^{2\pi} \int_0^{\infty} \int_0^{\infty} (f_l' f_j' - f_l f_j)\, gb \, \mathrm{d}\varphi \, \mathrm{d}b \, \mathrm{d}v_j \tag{2.13}$$

where n_j is the number density of the particle species j, f_j, f_j' its distribution function before and after the collision respectively, $g = |\boldsymbol{v}_l - \boldsymbol{v}_j|$ the relative velocity, b the impact parameter and φ the angle between the orbital plane and the plane containing the velocities of the two particles before collision. Sometimes (see, e.g., [Schi60]), this expression has been applied also to calculate the change in f produced by encounters with ions which inevitably involves the consideration of the COULOMB logarithm. Indeed, following the CHAPMAN-ENSKOG scheme [CC52] $(\delta f_\mathrm{e}/\delta t)_\mathrm{i}^\mathrm{B}$ can be written as

$$\left(\frac{\delta f_\mathrm{e}}{\delta t}\right)_\mathrm{i}^{\mathrm{B}} = - f_1 v n_\mathrm{i} Q_\mathrm{i}(v) \tag{2.14}$$

where f_1 is the second term of the series expansion for the distribution function and $Q_\mathrm{i}(v)$ the momentum transfer cross-section for $e - i$ collisions. If a pure COULOMB potential is assumed, the cross-section is found to diverge logarithmically in the impact parameter. To avoid this divergence, it has been suggested (see, e.g., [Lib59, MMS67, WW69]) to take into account plasma screening effects in the calculation of $Q_\mathrm{i}(v)$ which is connected with a delicate mathematical procedure, however. Without becoming involved in a general debate concerning this subject matter — a survey will be given in section 2.2.3.1. — we note that it has been favoured in the literature to cut off the integration in (2.13) with an upper limit b_0. Unfortunately, there is no general idea concerning the value that should be used for b_0 when an ionized gas is considered. In ideal plasmas a widely used approach is $b_0 = D$, whereas in nonideal ones this quantity is under discussion.

If charged particle interactions prevail as it is in highly ionized plasmas the preponderant number of collisions results in small deflections which in turn produce small changes in the velocities of the colliding particles. The occasional large deflections are entirely negligible and the value of $(\delta f_l/\delta t)_j$ will depend on the first and second derivatives of f_l rather than on f_l itself [Chan43]. Usually it is supposed that the BOLTZMANN collision term is not well suited to handle this type of interaction and the FOKKER-PLANCK term is preferred (see, e.g., [Des63]):

$$\left(\frac{\delta f_l}{\delta t}\right)_j^{\mathrm{FP}} = -\frac{\partial}{\partial \boldsymbol{v}}(f_l\langle \Delta(\boldsymbol{v})\rangle) + \frac{1}{2}\frac{\partial}{\partial \boldsymbol{v}}\cdot\frac{\partial}{\partial \boldsymbol{v}} : \ (f_l\langle \Delta\boldsymbol{v}\ \Delta\boldsymbol{v}\rangle) \tag{2.15}$$

where $\Delta\boldsymbol{v} = \boldsymbol{v}_l' - \boldsymbol{v}_l$ and, for the quantity A, $\langle A\rangle$ is defined by

$$\langle A\rangle = \int_0^\infty g f_j \,\mathrm{d}v_j \int_0^{2\pi} \mathrm{d}\varphi \int_0^\infty A b \,\mathrm{d}b\,. \tag{2.16}$$

If $\Lambda^{*2} \gg 1$ where $\Lambda^* = (4\pi\varepsilon_0 m_e v^2 b_0/e^2)$, an equivalent form to (2.16) reads (ROSENBLUTH and co-workers [RDJ57]):

$$\begin{aligned} \langle \Delta\boldsymbol{v}\rangle &= \frac{e^4}{4\pi\varepsilon_0^2 m_e^2}\ln \Lambda^* \frac{\partial}{\partial \boldsymbol{v}}\left(\frac{1}{v}\right) \\ \langle \Delta\boldsymbol{v}\ \Delta\boldsymbol{v}\rangle &= \frac{e^4}{4\pi\varepsilon_0^2 m_e^2}\ln \Lambda^* \frac{\partial}{\partial \boldsymbol{v}}\frac{\partial}{\partial \boldsymbol{v}} v\,. \end{aligned} \tag{2.17}$$

Clearly, if a pure COULOMB potential is used in computing the FOKKER-PLANCK coefficients the integration in (2.16) must be terminated also at some maximum to avoid the difficulties mentioned before.

To obtain a solution to the BOLTZMANN transport equation, a common procedure is to expand the distribution function in vector spherical harmonics and to truncate it after the first two terms. So, following the scheme which is in general use one looks for a solution of the form

$$f = f_0\left\{1 + \frac{\boldsymbol{v}}{v}\psi(v)\right\} \tag{2.18}$$

where f_0 is the MAXWELLIAN velocity distribution and $\psi(v)$ a function to be determined. Different methods were used in finding a solution for $\psi(v)$. SPITZER and co-workers [CSR50, SH52] derived an integro-differential equation for ψ which must be integrated numerically because the convergence of the expansion method used was unknown. Another way to attain a solution is to express ψ by a finite number of terms of a polynomial expansion in the particle velocities. In choosing the most economical form it is usual (see, e.g., [Schi60, SJB66]) to employ the LAGUERRE-SONINE polynomials of the order 3/2. In that way ψ may be expressed as follows

$$\psi(v) = \sum_\nu q_\nu S_\nu^{(3/2)}(z^2) \tag{2.19}$$

where z is a dimensionless velocity variable defined by $z = (v^2 m_e/2kT)^{1/2}$ and q_ν are the expansion coefficients. Taking into consideration the orthogonality and normalization properties of SONINE polynomials, when solving the BOLTZMANN transport equation one obtains a set of algebraic equations for the coefficients q_ν. However, only the first two of them are required to calculate the transport properties. In that way, the problem of finding the transport coefficients from kinetic equations is attributed to the determination of the expansion coefficients. The convergence of the method depends on the number of terms included in the polynomial expansion and will be discussed in particular case.

Different types of collision terms have been applied in deriving electron transport coefficients from the kinetic equation (2.12). SCHIRMER and FRIEDRICH [Schi60, SF58a, SF58b] actually calculated transport properties of partially ionized gases using the BOLTZMANN form of the collision term. SPITZER et al. [CSR50, SH52], on the other hand, applied the FOKKER-PLANCK collision term to compute the electrical conductivity of fully ionized gases, whereas SHKAROVSKY et al. [SJB66] carried out calculations using a combination of both collision terms: the FOKKER-PLANCK form for charged particle collisions and the BOLTZMANN collision term for the encounters between electrons and neutral particles. However, it should be emphasized that under similar expansions almost entirely identical results were obtained provided that the same grazing collision approximation is made. Indeed, using for the FOKKER-PLANCK coefficients the expression suggested by ROSENBLUTH et al. [RDJ57] it is simple to demonstrate analytically the equivalence between $(\delta f_e/\delta t)_i^B$ and $(\delta f_e/\delta t)_i^{FP}$. Obviously, the two approaches lead to similar results because the mathematical description of the cumulative effect of many simultaneous weak deflections is identical with the description of the many body problem in terms of many successive binary collisions (see also [Gra61]). Hence, the choice of the collision term does not influence the result of calculations, and all transport coefficients which emerge from the BOLTZMANN collision integral will be found to be the same as those derived from the FOKKER-PLANCK form.

It should be noted in this connection that the transport coefficients calculated in [CSR50, Schi60, SF58a, SF58b, SH52, SJB66] contain only terms which are of order $\ln \Lambda$ ($\Lambda = 12\pi\varepsilon_0 DkT/e^2$). A variety of subdominant (nonlogarithmic) contributions due to effects which are usually neglected in treatments of ideal plasmas is to be found in more precise expressions of transport coefficients. In particular, some of these effects are: (i) plasma microfield distortions by the colliding particle resulting in a dynamically instead of a statically screened potential, (ii) quantum diffraction effects, and (iii) exchange scattering. The contribution of these effects to the electrical conductivity in low-density fully ionized plasmas have already been analyzed by WILLIAMS and DEWITT [WW69] using the following representation for the collision integral

$$\frac{\delta f}{\delta t} = \left(\frac{\delta f}{\delta t}\right)_{SS} + \left(\frac{\delta f}{\delta t}\right)_{BD} - \left(\frac{\delta f}{\delta t}\right)_{BS} . \tag{2.20}$$

The first term involves the cross-section associated with the statically screened potential, whereas the second and third term involve the BORN approximation cross-section for the dynamically and statically screened potential, respectively. (Similar approaches for the collision term have been applied in [Kli75, REK80]). However, in their calculations, WILLIAMS and DEWITT, using the LIBOFF asymptotic form [Lib59] of $(\delta f/\delta t)_{SS}$ in the low-temperature limit, obtained a term $\ln(\Lambda) - 1.213$ in the denominator of the expression for the conductivity leading to the divergence indicated in figure 1.4. It is therefore difficult to relate the subdominant nonlogarithmic contributions due to the effects (i)—(iii) to the peculiarities observed in weakly nonideal plasmas.

Recently, the effect of the first term in (2.20) has been studied by ROGERS et al. [RWB81] without using the classical limit results of LIBOFF. The DEBYE screened cross-sections for electrons scattering from finite sized ions, neutrals, and other electrons were calculated from their correct quantum-mechanical expressions using the analytic effective potentials developed by ROGERS [Rog80]. The calculations correct therefore the major shortcomings of the original WILLIAMS-DEWITT treatment leading to improved agreement with experiments, particularly in strongly nonideal plasmas.

2.2.2. The Electric Transport Coefficients

In this section we deal with the transport coefficients of partially ionized gases following the treatment of SCHIRMER and FRIEDRICH [Schi60, SF58a, SF58b]. With regard to the numerical results it corresponds completely to the method of SHKAROFSKY [SJB66] and, if the particular case of a fully ionized gas is considered, to the method described by SPITZER [CSR50, SH52].

As mentioned before, the basic equation used by SCHIRMER and FRIEDRICH was from BOLTZMANN's type with BOLTZMANN collision terms both for charge-neutral and charge-charge particle interactions. The solution to the kinetic equation was expanded in terms of a complete set of SONINE polynomials. Specifically, for the electrical conductivity the following representation was found

$$\sigma = \left(\frac{e^2 n_e}{m_e}\right) q_0 \tag{2.21}$$

where q_0 is the coefficient of the lowest order polynomial in the expansion of the solution. Technically, q_0 must be determined from an infinite set of algebraic equations, but various approximate solutions can be obtained by truncating the expansion at given points. If n designates the level of approximation employed, the coefficient q_0 may be written in the form

$$q_0^{(n)} = \frac{3}{2} \frac{1}{w n_a} \cdot \frac{\Delta_{00}^{(n)}}{\Delta^{(n)}} \tag{2.22}$$

where

$$w = \left(\frac{2kT}{m_e}\right)^{1/2}.$$

$\Delta^{(n)}$ is a symmetrical determinant defined by

$$\Delta^{(n)} = \begin{vmatrix} I_{00} & I_{10} & \cdots & I_{n-1,0} \\ I_{10} & I_{11} & \cdots & I_{n-1,1} \\ \cdots & \cdots & \cdots & \cdots \\ \cdots & \cdots & \cdots & \cdots \\ I_{n-1,0} & I_{n-1,1} & \cdots & I_{n-1,n-1} \end{vmatrix} \tag{2.24}$$

and $\Delta_{00}^{(n)}$ is the minor of I_{00} in $\Delta^{(n)}$. Truncating after one and two terms (first and second approximation) yields, respectively,

$$\left.\begin{aligned} \sigma^{(1)} &= \frac{3}{2}\frac{e^2 n_e}{m_e}\cdot\frac{1}{w n_a} I_{00}^{-1}\,, \\ \sigma^{(2)} &= \sigma^{(1)}\left(1 - I_{10}^2/I_{00}\,I_{11}\right)^{-1}. \end{aligned}\right\} \tag{2.25}$$

In the case of a partially (one-fold) ionized gas the elements of $\Delta^{(n)}$ consist of a sum of three terms representing the contributions of $e - a$, $e - i$ and $e - e$ encounters to the collision term, respectively

$$I_{jk} = I_{jk}^{a} + \frac{n_i}{n_a} I_{jk}^{i} + \frac{n_e}{n_a} I_{jk}^{e}\,. \tag{2.26}$$

If additional particle species are present in the plasma, they can easily be taken into account by adding the corresponding terms in (2.26). Relating to the first two, the following representation can be used

$$I_{jk}^{a,i} = \sum_{l=1}^{j+k+1} A_{jkl} \int_0^\infty Q_{a,i}(z)\, z^{2l+1} F(z)\,\mathrm{d}z \tag{2.27}$$

where $F(z)$ is the velocity distribution given by

$$F(z) = \frac{4}{\sqrt{\pi}} z^2 \exp(-z^2)\,, \qquad z = \frac{v}{w}$$

and $Q_{a,i}(z)$ stands for the velocity dependent transport cross-section of the atoms and ions, respectively (concerning the explicit expressions used for them in actual calculations, see the next section).[1] The coefficients A_{jkl} are well defined numbers which can be found from the finite power series

$$S_j^{(3/2)}(z^2)\, S_k^{(3/2)}(z^2) = \sum_{l=1}^{j+k+1} A_{jkl} z^{2(l-1)} \tag{2.29}$$

where the SONINE polynomials of the order 3/2 are defined as

$$S_\nu^{(3/2)}(z^2) = \sum_{i=0}^{\nu} \frac{(-1)^i\,\Gamma(\nu + 5/2)}{i!\,(\nu - i)!\,\Gamma(i + 5/2)} z^i\,. \tag{2.30}$$

By introducing an electron collision cross-section in the form

$$Q_e^{(m)}(s) = 2\pi \int_0^\infty (1 - \cos^{2m}\chi)\, b\,\mathrm{d}b\,, \qquad m = 1, 2, \dots \tag{2.31}$$

where $s = g/(\sqrt{2}w)$ and g is the relative velocity of the colliding electrons, the third term in (2.26) may be written

$$I_{jk}^{e} = \sum_{\substack{l \\ i=2m}} B_{jkli} \sqrt{2} \int_0^\infty Q_e^{(m)}(s)\, s^{2l+1} F(s)\,\mathrm{d}s\,. \tag{2.32}$$

The quantities B_{jkli} can be obtained by comparison the corresponding coefficients of

[1] Apart from a numerical factor the definition of the I_{jk} agrees with that of the usual $\Omega^{(l)}$ integrals given by CHAPMAN and ENSKOG [CC52].

the following expansion

$$\sum_{li} B_{jkli}x^{2l}y^i \equiv 2^{-(j+k-1/2)} \sum_{p=0}^{\min(j,k)} \sum_{q=0}^{p} \sum_{r=0}^{p-q} (-1)^q\, 2^{p-q} \cdot \tag{2.33}$$

$$\cdot \left[\frac{3}{2} + p + x^2 y\right] \frac{\left(\frac{3}{2} + j + k + q - p\right) \dots \left(p + \frac{5}{2}\right) (j + k + 2q - 2p)\,!\, x^{2p} y^r}{q!\,(j + q - p)\,!\,(k + q - p)\,!\,(j + k + q - 2p)\, r!\,(p - q - r)\,!}$$

Both the A_{jkl} and the B_{jkli} are listed in table 2.1 up to the fifth order approximation. From the last relation follows that for $k = 0$ only coefficients with $i = 0$ and $i = 1$ appear in the expansion. In (2.32), however, these terms do not occur implying $I^{\mathrm{e}}_{j0} = 0$ for all values of j. Hence, electron-electron collisions have an effect on the electrical conductivity only in the higher order approximations ($n > 1$), whereas in the first one which closely resembles the expressions (2.8) and (2.9) they do not occur.

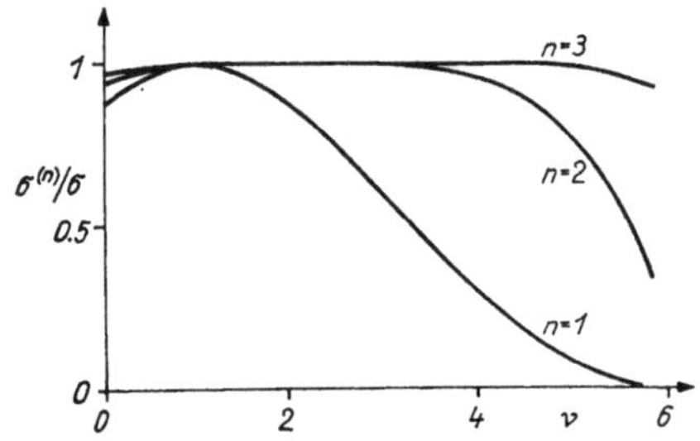

Fig. 2.2. Convergence of the expansions according to SCHIRMER and FRIEDRICH [SF58b] for a LORENTZ gas

The convergence of the procedure depends on the power of the velocity dependent transport cross-sections. Assuming $Q(v) = \mathrm{const} \cdot v^{-\nu}$, SCHIRMER and FRIEDRICH have investigated the convergence of the expansions in the range $-4 < \nu < 6$ neglecting the electron-electron interaction in their expressions (LORENTZ-gas). In the region $\nu > 0$ — because of $Q_{\mathrm{i}}(v) \sim \ln(1 + \Lambda^{*2})/v^4$ a completely ionized gas, for instance, is characterised by $\nu \approx 4$ — the expansion converges rapidly to the exact value computed by using complete SONINE polynomials (see also figure 2.2). However, if electron-atom interactions prevail as it is at lower degree of ionization the convergence deteriorates and a higher order approximation must be used. Particularly, this effect will become important in the case of heavy noble gases where the RAMSAUER minimum in the electron-atom collision cross-section has been observed.

If in the preceeding expressions the effect of neutral particles is neglected ($n_{\mathrm{a}} = 0$), the electrical conductivity of a completely one-fold ionized plasma in n-th order approximation is obtained. Assuming Coulomb potentials with a cut-off at a suitable distance b_0 both for electron-ion and electron-electron interactions, the remaining integrals can be solved analytically and the related approximation for the electrical conductivity appears as [RG75]

$$\sigma^{(n)} = \frac{6(\pi\varepsilon_0)^2\,(kT)^{3/2}}{e^2\sqrt{2\pi m_e}} \sum_i p_i^{(n)}\,(\mu^*)^i \tag{2.34}$$

where p_i are coefficients which remain of the same order of magnitude with rising i and μ^* is defined by

$$\mu^* = \frac{\mu}{2} \cdot \frac{D}{b_0} \tag{2.35}$$

μ being the plasma parameter and D the Debye length. For the plasmas of current interest μ^* is of the order of 10^{-1} so that the summation in (2.34) can be taken only over the first two terms. Expressing them owing to the rapid convergence of $\sigma^{(n)}$ in second order approximation, the explicit forms of the coefficients are [RG75]:

$$p_0^{(2)} = \frac{1 + 1.17 \ln \mu^*}{0.15 - 0.14 \ln \mu^* - 0.3 \ln^2 \mu^*}, \qquad p_1^{(2)} = \frac{1 + p_0^{(2)} (0.27 \ln \mu^* - 0.14)}{0.06 (\ln \mu^* - 1) + 0.12 \ln^2 \mu^*}. \tag{2.36}$$

The expressions (2.34) and (2.36) represent the counterpart to the well-known SPITZER relationship [SH52]

$$\sigma_{\text{SP}} = \frac{2(2kT)^{3/2}}{\pi^{3/2} m_e^{1/2}} \cdot \frac{(4\pi\varepsilon_0)^2}{e^2} \cdot \frac{\gamma_E}{\ln (3/\mu)}; \qquad \gamma_E = 0.5816 \tag{2.37}$$

frequently used in comparing experimental data with ideal plasma calculations. If we identify the cut-off distance with the DEBYE shielding length, that is setting $b_0 = D$ in (2.35) both conductivity formulas are found to correspond within 0.5% confirming the equivalence of the different collision terms as discussed in section 2.2.1. It should be noted in this connection that (2.35) is no longer a unique function of μ, unless the cut-off distance is expressed in terms of the DEBYE length.

In describing the kinetic properties of ionized gases, sometimes (see, e.g., [HM68, SJB66]) the transport coefficients are expressed in terms of an effective collision frequency. It is obtained by averaging the electron collision frequency

$$\nu_{ej} = n_j v Q_j(v) \tag{2.38}$$

in the following way [SJB66]

$$\langle \nu_{ej} \rangle = -\frac{4\pi}{3n_e} \int_0^\infty \nu_{ej} v^3 \frac{\partial f_0}{\partial v} \, dv \tag{2.39}$$

where f_0 is the MAXWELLian velocity distribution of electrons and n_j is the number density of the particle species j. With the use of effective collision frequencies, the electrical conductivity can directly be computed from

$$\sigma = \frac{e^2 n_e}{m_e} \frac{1}{\langle \nu_{ei} + \nu_{ea} \rangle g_\sigma} \tag{2.40}$$

where ν_{ei} and ν_{ea} are the electron-ion and electron-atom collision frequencies, respectively, and g_σ is the kinetic function according to SHKAROFSKY [SJB66] which is related to certain $e - i$, $e - a$ and $e - e$ collision matrix elements. By using the more explicit form of (2.39)

$$\langle \nu_{ej} \rangle = \frac{8}{3\sqrt{\pi}} n_j \left(\frac{m_e}{2kT}\right)^{5/2} \int_0^\infty v^5 Q_j(v) \exp\left(-\frac{m_e v^2}{2kT}\right) dv \tag{2.41}$$

one realizes that the first approximation in the treatment of SCHIRMER and FRIEDRICH corresponds to the above expression supposing $g_\sigma = 1$.[1] The higher approximations

[1] In the particular case of constant collision frequencies, the first approximation is identical with the exact solution computed using complete SONINE polynomials.

Table 2.1. The coefficients A_{jkl} and B_{jkli} according to SCHIRMER and FRIEDRICH [SF58a]

j	k	A_{jk1}	B_{jk12}	A_{jk2}	B_{jk22}	A_{jk3}	B_{jk32}	A_{jk4}	B_{jk42}	B_{jk44}
0	0	1	0	—	—	—	—	—	—	—
1	0	2.5	0	−1	0	—	—	—	—	—
1	1	6.25	0	−5	0.5	1	—	—	—	—
2	0	4.375	0	−3.5	0	0.5	0	—	—	—
2	1	10.938	0	−13.125	0.875	4.75	−0.25	−0.5	—	—
2	2	19.141	0	−30.625	2.4063	16.625	−0.875	−3.5	0.125	—
3	0	6.5625	0	−7.875	0	2.25	0	−0.1667	0	—
3	1	16.406	0	−26.25	0.9844	13.5	−0.5625	−2.6667	0.0625	—
3	2	28.711	0	−57.422	3.6914	40.688	−2.0391	−12.542	0.3906	—
3	3	43.066	0	−103.36	7.1060	91.547	−4.7461	−37.625	1.2227	0.0208
4	0	9.0234	0	−14.438	0	6.1875	0	−0.9167	0	0
4	1	22.559	0	−45.117	0.9023	29.906	−0.7734	−8.4792	0.1719	—
4	2	39.478	0	−94.746	4.2861	82.113	−3.3516	−32.885	0.8594	—
4	3	59.216	0	−165.81	9.8976	174.60	−8.3386	−88.730	2.6533	0.0573
4	4	81.422	0	−260.55	16.038	320.11	−15.553	−195.20	5.8988	0.2005

Table 2.1. continued

j	k	A_{jk5}	B_{jk52}	B_{jk54}	A_{jk6}	B_{jk62}	B_{jk64}	A_{jk7}	B_{jk72}	A_{jk8}	B_{jk82}	A_{jk8}
0	0	—	—	—	—	—	—	—	—	—	—	—
1	0	—	—	—	—	—	—	—	—	—	—	—
1	1	—	—	—	—	—	—	—	—	—	—	—
2	0	—	—	—	—	—		—	—	—	—	—
2	1	—	—	—	—	—	—	—	—	—	—	—
2	2	0.25	—	—	—	—	—	—	—	—	—	—
3	0	—	—	—	—	—	—	—	—	—	—	—
3	1	0.1667	—	—	—	—	—	—	—	—		
3	2	1.7083	−0.0313	—	−0.0833	—	—	—	—	—	—	—
3	3	7.6875	−0.1406	—	−0.75	0.0078	—	0.0278	—	—	—	—
4	0	0.0417	0	—	—	—	—	—	—	—	—	—
4	1	1.0208	−0.0104	—	−0.0417	—	—	—	—	—	—	—
4	2	6.4844	−0.1042	—	−0.6042	0.0052	—	0.0208	—	—	—	—
4	3	23.820	−0.4121	−0.0104	−3.4218	0.0332	—	0.2465	−0.0013	−0.0069	—	—
4	4	65.506	−1.1279	−0.0573	−12.547	0.1206	0.0052	1.3559	−0.0072	−0.0764	0.0002	0.0017

can be expressed through deviations of g_σ from unity differing only in the form of representation from the formalism described by SCHIRMER.

If a temperature gradient is present in the plasma which in general, will also cause gradients of concentration, there is an additional charge transport due to the tendency of the electrons to diffuse in direction of these gradients. In most cases, however, these contributions can be neglected because the corresponding electron fluxes are small compared to those produced by the external electric field. To demonstrate this we consider a partially ionized plasma possessing a temperature gradient ∇T parallel to the electric field $\boldsymbol{E}$. Let $\boldsymbol{j}_{\mathrm{E}}$ and $\boldsymbol{j}_{\nabla\mathrm{T}}$ be the current per unit area due to the electric field and the temperature gradient, respectively, then according to [Schi60] the ratio $|\boldsymbol{j}_{\nabla\mathrm{T}}|/|\boldsymbol{j}_{\mathrm{E}}|$ can be written

$$\frac{|\boldsymbol{j}_{\nabla\mathrm{T}}|}{|\boldsymbol{j}_{\mathrm{E}}|} = \frac{k\,|\nabla T|}{eE}\left\{(1-\alpha)\left(\frac{5}{4}+\frac{\chi_i}{2kT}\right)+\frac{5}{2}\,\frac{\Delta_{10}^{(n)}}{\Delta_{00}^{(n)}}\right\} \tag{2.42}$$

where α is the degree of ionization, χ_i is the ionization energy, $\Delta_{10}^{(n)}$ and $\Delta_{00}^{(n)}$ are minors of the corresponding elements in $\Delta^{(n)}$, and E is the magnitude of $\boldsymbol{E}$. To get an estimate for the above ratio we neglect α in (2.42), and for the quotient $\Delta_{10}^{(n)}/\Delta_{00}^{(n)}$ we use the result obtained for a LORENTZ gas. If the degree of ionization is not too low this quantity becomes approximately 3/5. Hence, for typical values $E = 10^4$ V/m and $T = 0.1\chi_i/k$ the following relation holds even if a gradient $\nabla T = 10^5$ K/m is assumed in the homogeneous region of the arc column

$$|\boldsymbol{j}_{\nabla\mathrm{T}}|/|\boldsymbol{j}_{\mathrm{E}}| < 10^{-2}\,. \tag{2.43}$$

2.2.3. Evaluation of Transport Cross-Sections

The most important problem in the calculation of electron transport properties is to take account of the mechanism of particle interactions. Within the frame of the binary collision model this can be done in terms of so-called transport or gas-kinetic cross-sections which can be written in the form

$$Q^{(l)}(v) = 2\pi\int_0^\pi (1-\cos^l\chi)\,\sigma(v,\chi)\sin\chi\,\mathrm{d}\chi\,. \tag{2.44}$$

Here χ is the deflection angle of the scattered particle and $\sigma(v,\chi)$ the differential cross-section for the scattering. In the classical limit, the transport cross-sections are most conveniently expressed as integrals over the impact parameter

$$Q^{(l)}(v) = 2\pi\int_0^\infty (1-\cos^l\chi)\,b\,\mathrm{d}b \tag{2.45}$$

where the angle of scatter for an arbitrary potential $U(r)$ is given by

$$\chi = \pi - 2b\int_{r_{\mathrm{m}}}^\infty\left(r^2\sqrt{1-b^2/r^2-2U(r)/m_e v^2}\right)^{-1}\mathrm{d}r \tag{2.46}$$

r_m being the minimal interparticle distance which ensures an elastic collision between two particles.

In quantum-mechanical treatment the expressions (2.44) can be integrated employing the recurrence relation for Legendre polynomials to give transport cross-sections in terms of phase shifts δ_k [Dev67]:

$$Q^{(1)} = \left(\frac{h}{\mu g}\right)^2 \sum_{k=0}^{\infty} (k+1) \sin^2 (\delta_{k+1} - \delta_k) , \tag{2.47a}$$

$$Q^{(2)} = \left(\frac{h}{\mu g}\right)^2 \sum_{k=0}^{\infty} \frac{(k+1)(k+2)}{2k+3} \sin^2 (\delta_{k+2} - \delta_k) , \tag{2.47b}$$

$$Q^{(3)} = \left(\frac{h}{\mu g}\right)^2 \sum_{k=0}^{\infty} \frac{k+1}{2k+5} \left\{\frac{(k+3)(k+2)}{2k+3} \sin^2 (\delta_{k+3} - \delta_k) + \right.$$
$$\left. + \frac{2(k^2+2k-1)}{2k-1} \sin^2 (\delta_{k+1} - \delta_k)\right\} \tag{2.47c}$$

where μ is the reduced mass for two particles and g is their relative velocity.

One realizes that the first of the above expressions is, in fact, the momentum transfer cross-section. The second expression is sometimes called viscosity cross-section, since it is important for this coefficient.

It is evident that, when using a kinetic theory based on collision considerations, the problem of the entire calculation of transport coefficients is reduced to the evaluation of the various scattering cross-sections which can be done both quantum-mechanically and classically using phase shifts or interaction potentials as basic information.

2.2.3.1. Charged Particle Interactions

In describing the electron-ion interaction it is most natural to assume the mutual potential energy as

$$U(r) = \frac{e_1 e_2}{4\pi\varepsilon_0 r} \tag{2.48}$$

where e_1 and e_2 are the charges of the two colliding particles. However, the employment of (2.48) in the calculation of electron transport properties is connected with serious difficulties because the related scattering cross-section is divergent. On the other hand, the plasma polarization effect leads to an excess of negatively charged particles (i.e. electrons) in the neighbourhood of an ion which, in turn, modifies the above potential distribution. This phenomenon was first discovered by Debye and Hückel [DH23] in investigating the properties of ionic solutions. Extending the concept of Debye shielding to dilute plasma conditions, the modified distribution takes the form of an exponentially shielded Coulomb potential,

$$U(r) = \frac{e_1 e_2}{4\pi\varepsilon_0 r} \exp(-r/D) , \tag{2.49}$$

where the characteristic distance describing the size of the shielding region is the Debye length D. Obviously it is this length at which in equilibrium the electrostatic forces are balanced by the charge cloud. Assuming that heavy charged particles have the same effect on the screening mechanism as electrons, D is given by (1.4). On the other hand, if the ions are assumed not to take part in the establishment of the shielding equilibrium, the related Debye length for a singly ionized gas appears as

$$D^* = \sqrt{2} D . \tag{2.50}$$

As demonstrated by HAHN et al. [HMMS72], shielding by electrons only can be considered to be an effective substitute for the dynamic screening effects that were taken into account, for instance, by WILLIAMS and DEWITT [WW69]. (Hence, this approach is sometimes called 'dynamic screening' while the former is usually designated as 'static screening'.) However, because the GOULD-DEWITT approximation [WW69] becomes divergent in the region of weak nonideality, it is difficult to verify more directly the proposed model by experiments. Recent investigations [GNPP78] indicate that with increasing n_D the effective screening radius approaches the value defined by (1.4), that is the model which includes shielding by electrons as well as heavy charged particles. Other authors [Lit75, Vi71], on the contrary, found a certain experimental proof for the validity of the effective dynamic screening model. In view of these circumstances the size of the effective shielding region is somewhat uncertain. Nevertheless, the potential distribution (2.49) can be treated as a model description for charged particle interactions in ionized gases.

In low-density (ideal) plasmas the shielding cloud is set up on a time scale very much shorter than the mean time between two collisions. Thus, it suggests itself to describe the process of momentum transfer by means of the binary collision integrals (2.26) taking into account the effect of the surrounding plasma through an adapted potential distribution. Unfortunately, when applying the exponentially screened potential (2.49), the corresponding transport cross-section cannot be completely expressed in terms of analytic functions. Hence, it has to be found numerically or approximations must be used leading to limitations of the analytic solutions to this problem.

Several calculations of transport cross-sections for screened potentials have been reported in the literature (see, e.g., [HMS71, Ki64, Lib59, MMS67, WW69]). LIBOFF [Lib59] split the interaction into a short range part with a pure COULOMB potential and a weak part governed by a screened potential, i.e.

$$U(r) = \frac{e_1 e_2}{4\pi\varepsilon_0} \begin{cases} r^{-1} & \text{for} \quad r < b_1 \,, \\ r^{-1} \exp\left(-r/D\right) & \text{for} \quad r \geqq b_1 \,. \end{cases} \tag{2.51}$$

Accordingly, the domain of integration in (2.45) was divided into two intervals: 0 to b_1 and b_1 to ∞. In evaluating the short range contributions to the transport cross-section the exact COULOMB expression for the scattering angle χ could be used. In the far region ($r > b_1$) the deflection angle was calculated substituting the screened potential in (2.46) together with the assumption that only small angle scatterings are relevant (KENNARD small-angle approximation [Ke38]). So, all the calculations could be done analytically resulting in the following expression for the momentum transfer cross-section

$$Q_i^l(v) = 4\pi \left(\frac{D}{\Lambda^*}\right)^2 (\ln \Lambda^* - 0.384) \tag{2.52}$$

where

$$\Lambda^* = 4\pi\varepsilon_0 m_e v^2 b_0 / e^2 \tag{2.53}$$

and $b_0 = D$. However, because the seperation parameter b_1 must satisfy the condition

$$L \ll b_1 \ll D$$

(L — LANDAU length) which only proves right above $\Lambda^* = 200$, the LIBOFF result has to be interpreted as the asymptotic high-energy limit not applicable to the low-temperature plasmas of current interest. Using the same arguments as LIBOFF a similar expression for the momentum transfer cross-section was also derived by KIHARA [Ki64].

Since it is difficult to choose a b_1 satisfying the above inequality even in ideal plasmas, MASON and co-workers [MMS67] improved the earlier calculations of LIBOFF and KIHARA dividing the range of integration into three intervals: 0 to b_1, b_1 to b_2 and b_2 to ∞. In the range $(0, b_1)$ and (b_2, ∞) the calculations were carried out analytically using the same arguments as before. The contribution to the cross-section from the range (b_1, b_2) was found by numerical integration and gave different results for the attractive and repulsive potential. So, as demonstrated in figure 2.3 the momentum

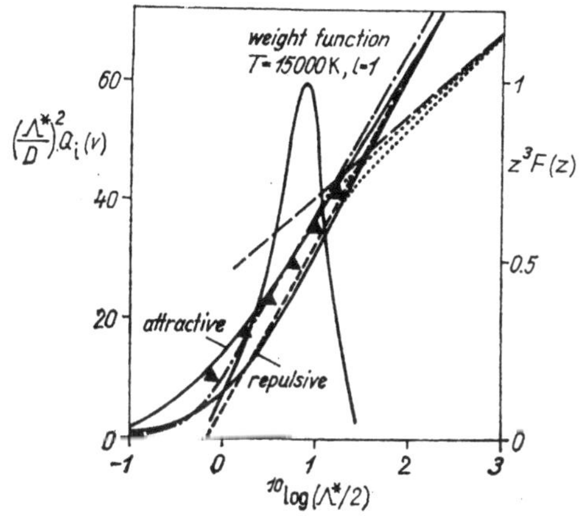

Fig. 2.3. Momentum transfer cross-sections for different potentials plotted against $^{10}\log\,(\Lambda^*/2)$; ——— repulsive and attractive screened potentials according to [MMS67], ... as before including quantum-mechanical corrections according to [HMS71], —·—·— COULOMB potential with a limitation of the impact parameter at $b_0 = D$, -------- the high-energy asymptotic form according to LIBOFF [Lib59], — — — screened potential in first BORN approximation, ▲ cut-off COULOMB potential according to SUCHY [Suc64]

transfer cross-section for the attractive potential proves to be larger in magnitude than the repulsive one. The LIBOFF asymptotic form (2.52) is obtained at high energies when $D \gg\gg L$ and the range of integration from b_1 to b_2 can be taken negligibly small.

From the physical point of view the numerically computed transport cross-sections [MMS67] represent the most satisfying expressions describing the mechanism of particle interactions in ideal plasmas. On the other hand, if one is looking for an analytical representation of the tables presented by MASON and co-workers, in a small energy range the numerical values for the attractive potential are found to be in close agreement with the standard COULOMB cross-section,

$$Q_i(v) = 2\pi \left(\frac{D}{\Lambda^*}\right)^2 \ln\,(1 + \Lambda^{*2})\,, \tag{2.54}$$

terminated at $b_0 = D$. Fortunately, in low-temperature plasmas this region corresponds almost entirely to the range where the effect of the weight function in the collision integrals (2.26) is large. To demonstrate this, in figure 2.3 the function $z^{2l+1}F(z)$ is shown in a normalized scale assuming a temperature $T = 15000$ K and $l = 1$. (In the higher approximations ($l = 2, 3$) the weight function will only slightly be shifted to the right.) So, the COULOMB cross-section (2.54) can be considered as a substitute for the numerical results [MMS67] resulting in a rather simple and exact treatment of charged particle interactions.

Applying the cut-off procedure in the evaluation of the electron collision cross-sections [(2.31), the first two of them necessary to calculate transport coefficients in higher approximations appear as [Schi60]

$$Q_e^{(1)}(g) = 16\pi \left(\frac{D}{\Lambda^*}\right)^2 \{\ln B - 1 + B^{-1}\} , \tag{2.55}$$

$$Q_e^{(2)}(g) = 32\pi \left(\frac{D}{\Lambda^*}\right)^2 \left\{\ln B - \frac{5}{2} + 3B^{-1} - 2B^{-2} + \frac{2}{3} B^{-3}\right\} \tag{2.56}$$

where

$$B = 1 + \Lambda^{*2}(g)/4 .$$

With the cross-sections (2.54)—(2.56) at hand recurrence relations can be used to give the I_{jk} in terms of trigonometric functions, the sine integral, and cosine integral. In [Schi60, SF58b] these expressions are listed up to the third order approximation which is sufficient for a large variety of temperatures and densities of interest.

When describing the kinetic properties in terms of collision frequencies then in such a cut-off model the effective electron-ion collision frequency has the form

$$\langle \nu_{ei} \rangle = \frac{4}{3} \sqrt{2\pi}\, n_i \left(\frac{e^2}{4\pi\varepsilon_0 kT}\right)^2 \left(\frac{kT}{m}\right)^{1/2} F(\Lambda) \tag{2.57a}$$

where

$$\begin{aligned} F(\Lambda) &= \sin \mu^* \left(\frac{\pi}{2} - Si\, \mu^*\right) - \cos \mu^* Ci \mu^* \\ &= \ln \left(\frac{1}{\mu^*}\right) - C + \frac{\pi}{2} \mu^* + 0(\mu^{*2}) , \qquad \mu^* = \frac{3}{2} \Lambda^{-1} \end{aligned} \tag{2.57b}$$

Si and Ci being the sine integral and cosine integral, respectively, and $C = 0.57722$.

We should mention here that the approximation of setting the interaction equal to zero for collisions with $b > b_0$ strictly speaking does not correspond to a reasonable assumption on the interaction potential. According to this procedure, only collisions with impact parameters smaller than b_0 contribute to the transport cross-section but the particle interaction will not be limited to a sphere with a radius b_0. Indeed, for these collisions the sphere of interaction is extended over the whole particle trajectory while collisions with impact parameters greater than b_0 will not be taken into account. Hence, a conceptually more satisfying approximation would be to use a potential which becomes zero outside a given sphere of interaction. Suchy [Suc64, Suc69] has calculated transport cross-sections as well as collision integrals taking for a basis a cut-off Coulomb potential of the form

$$U(r) = \frac{e_1 e_2}{4\pi\varepsilon_0} \begin{cases} r^{-1} - a^{-1} & \text{for} \quad r \leqq a , \\ 0 & \text{for} \quad r > a . \end{cases} \tag{2.58}$$

The momentum transfer cross-section $Q_{i,\,cut}(v)$ was found to be [Suc64]

$$Q_{i,\,cut}(v) = 2\pi \left(\frac{a}{\Lambda^*}\right)^2 \left\{\left(\frac{\Lambda^* \pm 1}{\Lambda^* \pm 2}\right)^2 \ln (1 \pm \Lambda^*)^2 - \frac{\Lambda^*}{\Lambda^* \pm 2}\right\} \tag{2.59}$$

where $b_0 = a$ in the expression for Λ^*. The upper and lower sign in (2.59) refer to the repulsive and attractive potential, respectively. At energies for which $\Lambda^* \gg 1$, (2.59) becomes asymptotically equal to

$$Q_{\mathrm{i,\,cut}}(v) = \left(1 - \frac{1}{2\ln\Lambda^*}\right) Q_i(v)$$

confirming the impact parameter approximation used in deriving the standard COULOMB cross-section. Obviously, this result reflects the dominance of small angle scatterings in which particle interactions take place in a small volume around the test particle cancelling the difference between the two treatments. In the opposite limit $\Lambda^* \ll 1$, the cross-section for scattering by a cut-off COULOMB force becomes one half of the result in (2.54). So, for the low-temperature plasmas situated between both extrema we should expect the two conceptions to differ slightly relating to the attenuation parameter in the interaction potential. The behaviour of the cross-section (2.59) for the attractive potential is illustrated in figure 2.3. As can be seen, it deviates from the standard COULOMB cross-section (2.54) only by some few per cent corresponding to a variation in the screening length of 10 ... 15%. Relating to the error involved in the experimental determination of this quantity, the measurements result, as a rule, in much larger uncertainties. Considering this fact, it seems to be justified to disregard the inconsistency of the impact parameter approximation and to use the more convenient expression (2.54) in subsequent calculations. SUCHY [Suc69] actually showed that when matching the results for a cut-off COULOMB potential (2.58) and an exponentially screened potential (2.49), the cut-off radius and the DEBYE length are related by the following expression

$$a = D(1.57 + 6/\Lambda)^{1/4} . \tag{2.60}$$

Accordingly, in ideal plasmas ($\Lambda \gg 1$) both lengths differ by a factor 1.1 which can easily be understood if we realize that the shielding distance in (2.49) in fact does not correspond to the actual range of this potential. In the region $7 < \Lambda < 50$ characteristic of weakly nonideal plasmas the difference is only slightly increasing confirming the conclusions reached before.

It remains to compare the classical momentum transfer cross-sections with those calculated by using quantum-mechanical methods. To this end, in figure 2.3 the numerically computed values [HMS71] for the attractive and repulsive screened COULOMB potentials are shown together with the corresponding first BORN asymptote. From the quantum-mechanical point of view these calculations are the most accurate available including the results of other authors (WILLIAMS and DEWITT [WW69], KIHARA [Ki64] and HONDA ([Hon64]) as a particular solution to this problem. As can be seen from the figure, in the region $\Lambda^* \geqq 10^3$ the cross-sections for the attractive and repulsive potentials are equal and agree with the first BORN asymptote accurate for long-range potentials at higher energies. The breakdown of the classical approximation begins above $\Lambda^* \approx 60$ outside the range of interest here. Moreover, at energies where quantum effects become apparent the cross-sections are too small in magnitude to cause any effect in the calculation of low-temperature plasma transport coefficients. This can be concluded when transforming the curves of figure 2.3 into a representation $Q_i(v)$ vs. ${}^{10}\log(\Lambda^*/2)$. For this purpose namely, we must multiply them by $(D/\Lambda^*)^2$ which is strongly decreasing with energy.

In [EK79, Kal70, SS73] the cross-section for a screened COULOMB potential in first BORN approximation was applied to evaluate electron transport coefficients of low-temperature plasmas. The authors obtained reasonable agreement with experiments in weakly nonideal plasmas tending to confirm the approaches used in their calculations. However, reference to figure 2.3 indicates that this agreement should not be taken seriously because the first Born asymptote fails in this region resulting in too large cross-sections. It will be shown below that the high pressure peculiarities associated with the process of momentum transfer can be related indeed to an enhancement of the interaction sphere but this is by reason of nonideal charged particle interactions rather than by quantum effects.

2.2.3.2. Interactions of Electrons with Neutral Particles

The presence of neutral particles in plasma gives rise to several problems:

First, the total collision integral for electrons becomes higher and therefore the absolute values of electron transport coefficients, especially electrical conductivity, become smaller.

Second, the problems connected with choosing the proper approximation to ensure the convergence of the Sonine polynomial expansion.

Third, a problem arising with the combination of the effects mentioned before is that the presence of neutral particles may have similar consequences as the violation of ideal-plasma assumptions. So, the eventual nonideality effects in dense partially ionized plasmas should be unfolded from overall experimental data by a detailed and careful analysis of the influence of neutral particles, taking concrete plasma conditions into account.

Several kinds of neutral particles may exist in plasma. Atoms in ground state are the most numerous. The ground states of hydrogen and heavy noble gas atoms are separated from the lowest excited states by an energy gap of approximately 10 eV or more. Therefore, the number density of the lowest excited configuration in low-temperature plasmas is typically four orders of magnitude smaller than the ground state number density. The total density of excited atoms does not exceed 10% of the ground state population in the most drastic cases.

However, the dimensions of excited atoms are much larger, so their properties should differ from the properties of ground state atoms. Particularly, their polarizabilities should have larger values and the momentum transfer with electrons should have larger cross-sections. So, the ground state and excited atoms could have quantitatively comparable effects on the electron transport cross-sections. By introducing the influence of excited atoms the total impact of electron-neutral particle interaction on the transport coefficients of partially ionized gases becomes larger than usually estimated [GLRP77].

Dissociation energies of molecules considered in this work are so low in comparison with the ionization energies of their atomic products that the neutral molecule number densities are several orders of magnitude lower than the populations of ground state atoms. They are so negligible that there exist no mechanisms which can make them important in the electron transport. An exception is, perhaps, the N_2 molecule which has a relatively high dissociation potential.

In the case of heavy noble gases, the electron-ground state atom collisions are characterized by the Ramsauer effect in which the cross-section passes through a minimum at low energy. This effect cannot properly be described by using a classical potential to represent the interaction. The most exact procedure would be to take the data for phase shifts and to compute cross-sections using the expressions (2.47a) to (2.47c). However, as DEVOTO [Dev67] demonstrated, to obtain transport coefficients it is sufficient to know the exact values for $Q_a^{(1)}$, that is for momentum transfer cross-sections. On the other hand, these are the only measurable quantities. Taking into consideration the problems concerning the proper description of the RAMSAUER effect we preferred to introduce averaged values of the cross-sections obtained from several experiments. In that way an empirical formula has been obtained which describes the electron-atom momentum transfer cross-section in the range of incident electron energy between 0.01 eV and 100 eV [PPV74]. This range is wide enough to give an accurate treatment of the transport coefficients of low-temperature ionized gases. The formula gives electron-ground state atom momentum transfer cross-sections in terms of the incident electron energy in the following way:

$$Q_a^{(1)}(E) = a \cdot \exp(-bE) + c \cdot E^d \cdot \exp(-fE) \tag{2.61}$$

where a, b, c, d, and f are coefficients determined numerically for several gases ([GPPR76, PPV74]). Their numerical values, listed in table 2.2, are chosen so that the cross-sections are obtained in 10^{-20} m^2 when E is in eV.

Table 2.2. Numerical values of the coefficients for ground state atom momentum transfer cross-sections[1]

Element	a	b	c	d	f
H	—	—	23.00	−0.56	0.02
Ne	—	—	1.49	0.30	−0.01
Ar	8.05	28.02	0.99	1.74	0.14
Kr	30.70	11.60	0.79	2.06	0.15
Xe	108.41	14.50	3.50	2.15	0.26

Considering a partially ionized gas in the condition of LTE, excited state densities appear to be small compared to ground state populations. This fact, combined with the lack of precise information concerning the interactions of electrons with excited atoms often led to conclusions about their negligible influence on the plasma transport properties.

There are very few experimental data concerning momentum transfer between electrons and excited atoms, so the necessary cross-sections for encounters of this type had to be estimated in order to investigate their eventual influence. Due to insufficient knowledge of necessary atomic parameters, quantum-mechanical calculations were practically impossible. For this reason, an estimate of the excited state cross-section was made based on the scattering of electrons on a polarization potential. The incident electron acts with its COULOMB field upon the atom which is then distorted and polarized. The so formed dipole is assumed to follow without retardation the moving electron with its orientation. The field of this dipole is acting on the incident electron by a polarization potential with an asymptotic form for the interaction energy (ideal dipole)

$$U(r) = -\frac{\alpha(n)\, e^2}{4\pi\varepsilon_0 r^4} \tag{2.62}$$

[1] In the case of hydrogen and xenon the fitting procedure has been repeated resulting in coefficients differing slightly from those reported in the original publications.

where $\alpha(n)$ is the polarizability in the state of quantum number n. The deflection angle for this scattering problem was calculated substituting the above potential distribution into (2.46). It can be represented in the following form

$$\chi = \left| \pi - \frac{2b}{r_1} K\left(\frac{r_2}{r_1}\right) \right| \tag{2.63}$$

where K is the complete elliptic integral of the first kind and r_1, r_2 are roots of the denominator in (2.46). A short analysis of the scattering process shows that the potential (2.62) allows a capture of the colliding electron if

$$b \leqq b_{\min} = \left(\frac{2\alpha(n)\, e^2}{\pi \varepsilon_0 m_e v^2}\right)^{1/4} . \tag{2.64}$$

Such collisions are treated as inelestic ones leading to total momentum loss ($\chi \equiv 90°$). On the basis of this model the scattering cross-section appears as

$$Q_{a,n}(v) = \pi b_{\min}^2 + 2\pi \int_{b_{\min}}^{\infty} (1 - \cos \chi)\, b\, db\, . \tag{2.65}$$

The integration in (2.65) was carried out numerically with the only modification that a cut-off b_0 was taken for the upper limit to account for screening effects by the surrounding plsama. In doing so the cross-section for momentum transfer between electrons and excited atoms was found to be of the following form

$$Q_{a,n}(v) = \begin{cases} \pi b_{\min}^2 \, (1 + g(b_0)) & \text{if} \quad b_{\min} < b_0\,, \\ \pi b_0^2 & \text{if} \quad b_{\min} \geqq b_0 \end{cases} \tag{2.66}$$

where $g(b_0)$ represents the effect of elastic collisions. Actually, $g(b_0)$ proved to be a rather weak function of the cutting length owing to the r^{-4} dependence of the polarization potential. It has thus been replaced by its asymptotic value $g(\infty) = 0.21$ in subsequent calculations.

In the case of hydrogen where exact atomic parameters are available the polarizabilities can be evaluated using the expression [Pop77]

Table 2.3. Polarizabilities of atomic hydrogen states

Quantum number	Ionization energy/eV	Statistical weight	α/m^3
1	13.60	2	$6.68 \cdot 10^{-31}$
2	3.40	8	$2.12 \cdot 10^{-29}$
3	1.51	18	$2.32 \cdot 10^{-28}$
4	0.85	32	$1.19 \cdot 10^{-27}$
5	0.55	50	$5.28 \cdot 10^{-27}$
6	0.38	72	$1.57 \cdot 10^{-26}$
7	0.28	98	$4.16 \cdot 10^{-26}$
8	0.21	128	$9.64 \cdot 10^{-26}$
9	0.17	162	$2.15 \cdot 10^{-25}$
10	0.14	200	$4.30 \cdot 10^{-25}$
11	0.12	242	$8.03 \cdot 10^{-25}$
12	0.10	288	$1.43 \cdot 10^{-24}$

Table 2.4. Polarizabilities of atomic argon states

Configuration	Average ionization energy/eV	Statistical weight	α/m³
$3p$	15.75	1	$1.673 \cdot 10^{-30}$
$3d$	1.68	60	$1.06 \cdot 10^{-28}$
$4s$, $4p$	3.31	48	$2.755 \cdot 10^{-27}$
$4d$, $4f$	0.85	144	$4.815 \cdot 10^{-27}$
$5s$, $5p$ ...	0.81	300	$8.55 \cdot 10^{-27}$
$6s$, $6p$ ...	0.46	432	$1.22 \cdot 10^{-26}$
$7s$, $7p$ ...	0.36	588	$2.58 \cdot 10^{-26}$
8 ...	0.23	768	$1.11 \cdot 10^{-25}$
9 ...	0.16	972	$3.8 \cdot 10^{-25}$

Table 2.5. Polarizabilities of atomic krypton states

Configuration	Average ionization energy/eV	Statistical weight	α/m³
$4p$	14.00	1	$2.51 \cdot 10^{-30}$
$4d$, $4f$	1.29	44	$1.08 \cdot 10^{-28}$
$5s$, $5p$	3.24	48	$3.2 \cdot 10^{-27}$
$5d$, $5f$	0.89	44	$2.1 \cdot 10^{-27}$
$6s$, $6p$	1.01	32	$1.45 \cdot 10^{-27}$
7 ...	0.54	88	$9.55 \cdot 10^{-27}$
8 ...	0.39	58	$2.51 \cdot 10^{-26}$
9 ...	0.31	72	$5.01 \cdot 10^{-26}$

Table 2.6. Polarizabilities of atomic xenon states

Configuration	Average ionization energy/eV	Statistical weight	α/m³
$5p$	12.13	1	$4.02 \cdot 10^{-30}$
$5d$, $5f$	1.83	144	$8.54 \cdot 10^{-28}$
$6p$, $6s$	3.15	48	$7.07 \cdot 10^{-27}$
$6d$, $6f$	0.95	144	$1.22 \cdot 10^{-26}$
7 ...	0.92	588	$1.73 \cdot 10^{-26}$
8 ...	0.55	768	$3.65 \cdot 10^{-26}$
9 ...	0.26	972	$1.18 \cdot 10^{-25}$

$$\alpha(n) = \frac{e^2}{4\pi^2\varepsilon_0 m_e} \sum_{\substack{k \\ k \neq n}}^{n^*} \frac{f_{kn}}{\nu_{kn}^2} + \int_{\nu^*}^{\infty} \frac{\partial f}{\partial \nu} \cdot \frac{d\nu}{\nu^2} \tag{2.67}$$

where n^* is defined as the last discrete level, f_{kn} are oscillator strengths of the corresponding bound-bound transitions, and the value of $\frac{\partial f}{\partial \nu}$ is to be obtained by extrapolating the function $f_{kn}(\nu)$ into the continuum; $\nu_{kn} = (E_k - E_n)/h$ where E_k and E_n are energy levels of the hydrogen atom and ν^* refers to the frequency labelled by n^*.

The polarizabilities of heavy noble gases have been estimated by POPOVIĆ [Pop77] using the procedure described by HIRSCHFELDER et al. [HCB54]. They are listed in tables 2.4–2.6.

The general behaviour of the cross-sections (2.66) is shown for hydrogen in figure 2.4 where they are presented for $n = 1, 2$ together with the corresponding ground state cross-section obtained from the empirical expression $Q_a^{(1)}$ corresponding to (2.61).

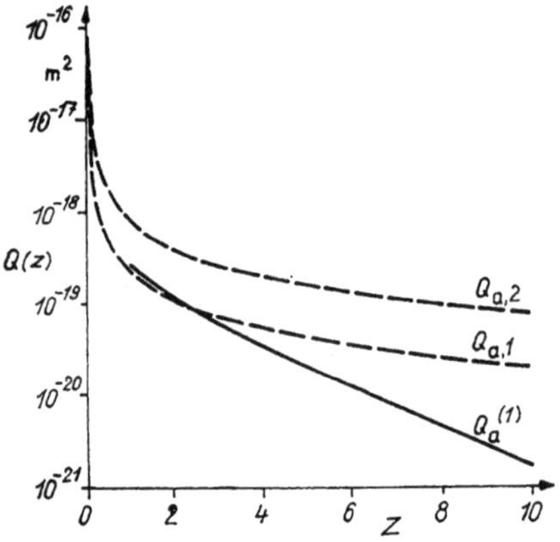

Fig. 2.4. Ground-state and excited-state momentum transfer cross-section for hydrogen $Q_{a,1}$ and $Q_{a,2}$ plotted against the dimensionless velocity z ($= \sqrt{m_e v^2/2kT}$ with $T = 15\,000$ K). Q_a^1: equation (2.61) representing experimental data

The cut-off value b_0 was assumed to be $5 \cdot 10^{-9}$ m which compares with the Debye length characteristic of weakly nonideal plasmas. Figure 2.4 suggests that the expressions (2.66) seem to have a correct functional form at least in the domain $z < 3$ where the effect of the weight function in the collision integrals is large. Therefore, they should be useful to estimate the influence of excited atoms on the transport properties of dense low-temperature plasmas.

With the excited state cross-sections at hand, the elements I_{jk} in (2.26) can be written as

$$I_{jk} = \frac{1}{U_0} \sum_n g_n \exp\left(-\frac{\chi_i - E_n}{kT}\right) I_{jk}^{a,n} + \frac{n_i}{n_a} I_{jk}^{i} + \frac{n_e}{n} I_{jk}^{e} \tag{2.68}$$

where g_n is the statistical weight of the level with the energy E_n and U_0 is the partition function of neutral particles. Technically, the cross-sections (2.66) increase indefinitely as a function of n and the summation in (2.68) has thus to be truncated at some n_{max}. This is necessary to avoid the growth of the cross-sections (2.66) beyond the ion transport cross-section because the highly excited atom can no longer be treated as a dipole with respect to the colliding electron.

Inclusion of excited atoms gives rise to higher values of the effective electron collision frequency, and so, to lower values of the electrical conductivity.

2.3. Non-Ideality Corrections to the Transport Coefficients

Before we shall deal with high-density corrections relevant to weakly nonideal plasmas, a short analysis of situations where the ideal plasma conditions are violated would be helpful.

As already mentioned in chapter 1.1., an ideal plasma is characterized by a large number of charged particles in the DEBYE sphere implying $D \gg r_m$ where r_m is the mean interparticle distance. In such conditions, D corresponds to the actual range of the interaction potential, and the great majority of collisions takes place at distances governed by the long-range part of the (more or less screened) COULOMB potential resulting in weak deflections of colliding particles. On the other hand, the less frequent close encounters particularly those in which particles approach each other at distances $r \leqq r_m$ are almost entirely negligible.[1] As regards the change of momentum produced when particles collide at distances $r_m < r < D$, it can properly be described in terms of binary collision cross-sections notwithstanding the collective nature of this interaction process. The reason is that the latter type of collision represents the limiting case of a stochastic process which can correctly be attributed to the additive sum of many successive binary collisions applying the expressions derived for two-body encounters to the range $r > r_m$. Accordingly, in ideal plasmas neither effects connected with short-range deviations from the pure COULOMB potential have to be taken into account nor is any approach in summing the many collective weak collisions necessary other than that of supposing a proper shielding mechanism to ensure the convergence of the transport collision integrals.

Turning to nonideal plasmas, the physical picture is quite different. First, the shielding equilibrium characteristic of ideal plasmas may hardly be expected since the local charge density at distances r from the collision centre

$$\varrho = \sum e_j n_j \exp\left\{-\frac{e_j\varphi(r)}{kT}\right\}$$

n_j and e_j being the average density and charge of the particle species j, respectively) an no longer be reduced to a linear expression in $\varphi(r)/kT$ leading to deviations from he static DEBYE-HÜCKEL potential (2.49) after all. Obviously, this difference is dependent largely on the shielding situation in the environment of the charge carriers and will be more pronounced as the number of charged particles in the DEBYE sphere decreases. If the interaction process has still the character of a collision as it is in weakly nonideal plasma the resultant non-ideality corrections may be obtained on the basis of collision considerations. In strongly nonideal plasmas this approach should be avoided because here particles must be considered as being in a continuous process of collision [NV79, VN77].[2]

Secondly, what becomes increasingly necessary as particle densities are raised is that the interaction potential should be modified for short distances depending on the special fine structure of the scattering ion under investigation. Indeed, in nonideal

[1] A rough estimate shows that for $\Lambda \gg 1$ the average loss of momentum due to close collisions ($\chi > 90°$) is only some few per cent of the total momentum loss including the many distant encounters too.

[2] Within the framework of the binary collision approximation, the ratio of the average time interval between two consecutive encounters and the collisions time proves to be of the order of $\tau_f/\tau_c \approx \lambda/L \approx \gamma^{-3}/\ln \Lambda$ where λ is the mean free path and L is the LANDAU length. Hence, $\tau_c \ll \tau_f$ even in the weakly nonideal region and consideration of transport processes in terms of collision cross-sections appears to be an acceptable way. However, it is also to be noted here that highly correlated particle states may be involved in the process of strong interaction the life-time of which is much greater than τ_c [VK77a, VK77b].

plasmas the most frequent collisions are the strong ones taking place at distances $r \lessapprox r_m$. Thereby, the particles may approach each other so close that neither the trajectory can be considered as classical nor can the specific features of the electronic shells be ignored any longer when studying the electron interaction with different ions. Clearly, non-ideality corrections connected with short-range deviations from the ideal point charge are the more important the more dense the plasma is. In section 2.3.1. the effect of these corrections on weakly nonideal plasmas will be estimated reducing the problem of the entire calculation of transport coefficients to the computation of the momentum transfer cross-sections.

Third, due to the strong correlation involved in the interaction between ions and low-energy continuum electrons quasi-bound particle states are in question in nonideal plasma leading to a modification of the free electron definition after all. In developing an approach to the calculation of weakly nonideal transport properties including highly correlated particles, different correction terms have been obtained ([Kh78, KN73, VK77b]). However, no preference can be given to any one of the proposed models because none of them does without using more or less justified assumptions concerning the specific features of particle motion.

Finally, there are some other effects which are expected to be of considerable importance in the strongly nonideal regime. They cannot be understood on the basis of corrections to the binary collision model and require a more consistent treatment within the frame of a quantum-statistical many-body theory. These studies are of much interest in so far as the weakly nonideal plasma constitutes the limiting case of the strongly nonideal one.

2.3.1. Effects of Short-Range Forces

Effects of short-range forces are mainly due to the following reasons:

(i) Deviations originally associated with quantum effects in the electron-ion interaction at distances $r \lessapprox \lambda_B$. Properly speaking these effects reflect the quantum-mechanical peculiarities of the radial distribution function which differs significantly from the corresponding classical one if the interaction takes place at distances $r \lesssim \lambda_B$.

(ii) Deviations associated with the specific features of the electron interaction with different ions for very near encounters.

The problem of including quantum-mechanical effects in treatments of nonideal plasmas has been the subject of many studies ([Kli75, ER79, NV79, RMKK79, BKM80, REK80, RWB81, MR82, Röp82]). A number of authors ([ER79, RMKK79, BKM80, REK80, MR82]) have formulated a theory of transport properties within a general approach to the quantum-mechanical many body problem. The interaction between charged particles was described by model potentials of the following form ([REK80])

$$V(r) = \frac{e_1 e_2}{4\pi\varepsilon_0}\left[\frac{1}{r} - \frac{A}{r}\exp(-Br) - C\exp(-Br)\right] \tag{2.69}$$

which includes the COULOMB potential ($A = C = 0$) as well as the HELLMANN potential ($C = 0$) as special cases. In particular, the electrical conductivity of an electron-proton gas was computed in this way, and there are at present efforts (see, e.g., [Röp82] and references therein) to extend these calculations to partially ionized plasmas.

EBELING and RÖPKE [ER79] actually have derived the following expression for the electrical conductivity of a completely ionized plasma

$$\sigma^* = \frac{1}{f} \cdot \frac{1 - Q}{L} \tag{2.70}$$

where f is a numerical factor which depends on the kind of expansion used in solving the kinetic equation. It will only be mentioned here that Q is the so-called relaxation function which is related to the DEBYE-ONSAGER relaxation effect and L is the generalized COULOMB logarithm (concerning the explicit expressions for L and Q, see below). To demonstrate the effects of both quantum corrections and short-range forces, Q and L have been calculated in [ER79] for H- and Cs-plasmas both classically and quantum-mechanically.[1] The quantum-mechanical effects decrease L and Q in comparison to their classical values. However, in the region $n_{\mathrm{D}} \approx 1$ the related corrections are small and do not exceed 1% both in the case of L and Q. Moreover, the cesium data approach the numerical values for hydrogen (at $n_{\mathrm{D}} \approx 1$ within 1%) which clearly shows that the COULOMB potential may be considered a model description for charged particle interactions in weakly nonideal plasmas.

An alternative calculation of plasma transport properties was done by two other groups. ROGERS et al. [RWB81], using a CHAPMAN-ENSKOG type solution to the quantum-mechanical kinetic equation of GOULD and DEWITT [GW67], have improved the previous calculations of WILLIAMS and DEWITT [WW69] taking into account the effect of neutral particles as well as electrons scattering from the electronic shells of the ions. The phase shifts required to evaluate the relevant transport cross-sections were calculated using the analytic effective potentials developed by ROGERS [Rog80]. These potentials have the form

$$V_{\mathrm{eff}}(r) = \frac{e_1 e_2}{4\pi\varepsilon_0 r} \left\{ Z \exp\left(-\frac{r}{D}\right) + \sum_1^{n^*} \nu_n \exp\left(-\alpha_n r\right) \right\} \tag{2.71}$$

where Z is the net ionic charge, n^* is the principal quantum number of the highest occupied shell, ν_n is the number of electrons in the shell labelled by n, and the α_n are screening constants chosen to reproduce the bound state energies of the ion with net charge $Z - 1$. If the effect of short-range cores and neutral scatters is neglected in (2.71), then $V_{\mathrm{eff}}(r)$ approaches the potential distribution (2.49), and the treatment of ROGERS et al. reduces to a quantum BOLTZMANN equation in which charged particles interact via a statically screened COULOMB potential. Reference to figure 2.3 shows that the corresponding quantum-mechanical cross-section has the peculiar property to behave classically in the low-temperature limit. Hence, in weakly nonideal plasmas the numerical results are expected to be close to the classical ones, and the essential improvement in the method of ROGERS et al. is to take account of short-range core corrections and neutral scatters.

NORMAN and VALUEV [VN77, NV79] have developed an approach which takes into account quantum effects within the framework of the classical molecular dynamics

[1] Parameter values for Cs: $A = 1.371$, $B = 9.64 \cdot 10^9\,\mathrm{m}^{-1}$, $C = 0$.

method [AW59]. The essence of this technique is a numerical integration of classical equations of motion. The difference between the real electron trajectory and the classical one is taken into account by changing the electron-ion interaction potential at small distances. It is appropriate to perform the potential replacement using the thermodynamical pseudopotentials because in such a way the equality of the classical and quantum-mechanical radial distribution function is achieved. In general, the thermodynamical pseudopotential $V^*(r, T)$ is defined as

$$\exp\left(-\frac{V^*(r, T)}{kT}\right) \equiv \lambda_B^3 \sum |\psi_n(r)|^2 \exp\left(-\frac{E_n}{kT}\right) \tag{2.72}$$

where E_n and ψ_n are energy levels and corresponding wave functions in the state n. The main feature of these potentials is that they are temperature dependent, with the peculiar property of being finite at $r = 0$.

The computation has been performed in [NV79] for a two-component electron-ion plasma, and the numerical results were presented in the form of reduced conductivities $\sigma^*(n_D)$. In the range $0.05 < n_D < 0.5$ the predictions lie near the bottom of the experimental data region and seem to have qualitatively correct behaviour as the plasma parameter increases. Unfortunately, the applicability of the method proposed in [NV79] is restricted to plasmas with $n_D < 1$ since the computational error rises with decreasing density.

To show more directly the effect of the thermodynamical pseudopotential in the region of weak non-ideality, a calculation of the electron-proton cross-section has been made using the two-particle effective potentials presented by Barker [Bar71]. Actually, the potential distributions [Bar71] were represented in the following approximative form

$$V^*(r, T) = \begin{cases} X(T)\, r - Y(T) & \text{if} \quad r < r^* \\ -\dfrac{e^2}{4\pi\varepsilon_0 r} & \text{if} \quad r \geqq r^* \end{cases} \tag{2.73}$$

where X, Y are temperature dependent coefficients and $r^*(X, Y)$ is the point at which both expressions of (2.73) intersect. The momentum transfer cross-section was obtained numerically substituting the above potential into (2.45). As a result, figure 2.5 shows the ratio $1 - Q^*(T)/Q_i(w)$ as a function of kT where $Q^*(T)$ refers to the cross-section computed from the potential (2.73) and $Q_i(w)$ is the standard Coulomb cross-section (2.54) for the particular velocity $w = (2kT/m)^{1/2}$. It can be seen that the non-Coulomb short-range corrections decrease the cross-section and that the effect becomes more pronounced as the temperature decreases. This feature reflects the fact that the classical radial distribution function does no longer adequately describe the probability of finding two oppositely charged particles in a chosen distance since at low temperatures an appreciable number of them are bound. However, from the results it can be seen that there is some justification for ignoring this difference because the effect does not exceed a few per cent for plasmas of current interest.

It was mentioned above that scatterings from the electronic shells of the ions may affect the transport properties, and in the method of Rogers et al. [RWB81] this effect was included through the effective potentials (2.71). Another approach to this problem is developed by Podlubny and Ragovski [Pod75, PR80] who derived short-range corrections to the quantum-mechanical cross-sections (2.47) by assuming the phase shifts to be of the following form

$$\delta_k = \eta_k + \varepsilon_k \tag{2.74}$$

where η_k are the Coulomb values and ε_k are corrections due to the effect of the short-range cores. Then, the momentum transfer cross-section (2.47a) can be written as

$$Q^{(1)}(v) = Q^{(1)}_{\delta_k=\eta_k} + \Delta Q^{(1)} \tag{2.75}$$

with

$$\Delta Q^{(1)} = \left(\frac{h}{\mu g}\right)^2 \sum_{k=0}^{\infty} (k+1)\,\{\sin^2(\eta_{k+1} + \varepsilon_{k+1} - \eta_k - \varepsilon_k) - \sin^2(\eta_{k+1} - \eta_k)\}\,.$$

Analytical expressions for $\Delta Q^{(1)}$ have been suggested in [Pod75] where ε_k was estimated on the basis of the quantum defect method. These calculations were improved later ([PR80]) by an expansion of the Herman-Skillman type [HS63] with the following principal results: the correction $\Delta Q^{(1)}$ may be positive as well as negative depending on the chemical element and the temperature range under consideration (see fig. 2.5). At low and moderate temperatures the relative value of this term as compared with the Coulomb cross-section is small which is in full agreement with the other computational results [ER79, RWB81].

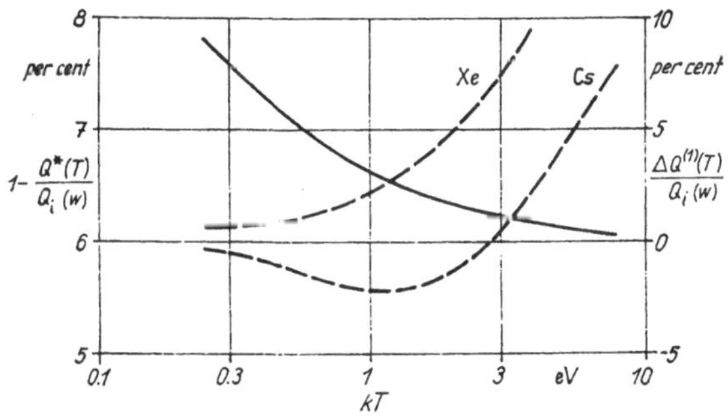

Fig. 2.5. Influence of short-range corrections on the electron-ion momentum transfer cross-section (the cutting length in $Q_i(w)$ was assumed to be 10^{-8} m), ——— the effect of pseudopotentials, — — — the effect of scatterings from the electronic shells according to [PR80]

2.3.2. Non-Debye Screening

If one is willing to accept the Debye shielding as an ideal model for charged particle interactions, then the basic hypothesis which is no longer fulfilled in nonideal plasmas is the assumption of a large number of charged particles in the Debye sphere. Under such conditions the interaction will not be limited to the range $r \lesssim D$ resulting in a macroscopic nonideal plasma conductivity. The related corrections should be identical for all plasmas with the same shielding conditions, i.e. the same non-ideality parameter γ or the same number of particles in the Debye sphere.

Many attempts to understand and explain the properties of nonideal plasmas have been made on the basis of corrections to the binary collision model. Among them a number of semiempirical methods have been devised which rely upon expressions for the transport coefficients containing an adjustable parameter to simulate nonideal behaviour.

The first approach of Bakeev and Rovinski [BR70b, BNRS71] used the Spitzer relationship [Sp56]

$$\sigma = \frac{aT^{3/2}}{\ln \Lambda} \tag{2.76}$$

to reproduce electrical conductivities derived from pulsed arc measurements on heavy noble gases. To fit (2.76) to the experimental data the authors proposed to modify the constant a, and instead of the original Spitzer value $a_{\mathrm{Sp}} = 1.55 \cdot 10^{-2}\ (\Omega \cdot \mathrm{m\ K}^{3/2})^{-1}$ they recommended $a = 8.8 \cdot 10^{-3}$ for argon, $a = 9.5 \cdot 10^{-3}$ for krypton and $a = 9.2 \cdot 10^{-3}$ for xenon in the same units as a_{Sp}. Unfortunately, these constants have no physical meaning and cannot be used to explain quantitatively such plasmas or to derive conclusions about theoretical models.

Another interpretation of that type of experimental results was given by Rovinski in [Rov72a, Rov72b]. The lower values of the electrical conductivity were assumed to be due to an enhancement of the effective ionic charge, and thus comparison was made with the Spitzer formula extended to completely ionized plasmas with variable ionic charges

$$\sigma = \frac{a_{\mathrm{Sp}}\, T^{3/2}}{Z_{\mathrm{eff}}^2 \ln \Lambda_1} \tag{2.77}$$

where $\Lambda_1 = \Lambda / Z_{\mathrm{eff}}$. The correction factor Z_{eff} gives rise to higher values of the electron-ion collision frequency, and accordingly, to lower conductivities. It was adjusted by Rovinski to reproduce a number of experiments including his own. The following representation was found to succesfully average the experimental data [Rov72a]

$$Z_{\mathrm{eff}}^2 = \exp\left(3.6 \cdot 10^3\ \mathrm{K\,eV}^{-1} \cdot \gamma \cdot \frac{\chi_i}{T}\right) \tag{2.78}$$

γ being the non-ideality parameter (1.2) and χ_i the ionization energy of the atom under consideration. One realizes that (2.78) indeed results in an effective single ion charge $Z_{\mathrm{eff}} > 1$ which tends to unity if the plasma becomes ideal ($\gamma \ll 1$). However it turns out that the proposed model is not sufficient to reproduce the true electrical conductivity over a wide range of parameters since effects of neutral particles which modify the temperature dependence of the conductivity function $\sigma(T)$ have been neglected in the treatment of Rovinski.

A consistently more accurate procedure to account for nonideal behaviour was described by Günther et al. [GR72, RG73, GPPR76, RG75, GLRP77, GLR83], Kulik et al. [Ku72, KKR72, KRR77], and Goldbach et al. [GNPP77, GNPP78] who used the correct $\ln \Lambda \gg 1$ expressions for the conductivity of partially ionized plasmas but considered the cut-off length entering the calculation of the ion transport cross-section as an adjustable parameter. The details of the procedure are described in section 5.1.

Günther and co-workers actually used for the free parameter

$$x = \frac{r_s}{D} \tag{2.79}$$

where D is the DEBYE length (1.4) and r_s an effective screening radius which should be adjusted to reproduce experimental data. The fitting procedure takes into account the incomplete screening of the scattering centre at distances $r \geqq D$ if n_D becomes comparable to unity. For such distances every ion should act like a simple point charge regardless of its chemical structure, and r_s should be a function of n_D only. The general behaviour of the measured function $x(n_D)$ is shown in figure 2.6. It suggests that the interaction sphere of the charge carriers must be enlarged owing to the incomplete screening conditions in nonideal plasmas.

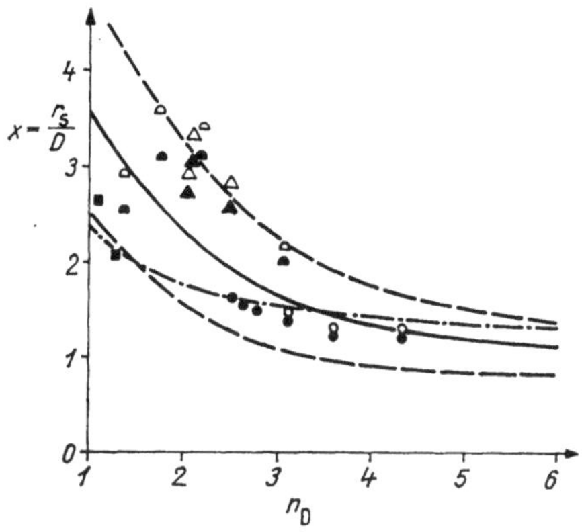

Fig. 2.6. Effective screening radius for the interaction of charged particles; full points: excited atoms included, open points: excited atoms as ground states, ● stationary argon arc [GNPP 78], ◓ pulsed argon arc [GLR 83], ■ pulsed xenon arc [GR72], ▲ pulsed hydrogen arc [GPPR76], solid curve: equation (2.80), dashed curves: error limits according to estimate equation (5.14), dot-dashed curve: equation (2.82)

The effect of the excited atoms can be seen by a comparison of the full and open marks in figure 2.6. The improved calculation including the cross-sections of the excited levels is represented by the full marks, whereas the open marks denote the traditional calculation. The effect can be rather important and depends on the gas species and, of course, on the plasma composition. It can be seen that essentially all the full points are within the experimental error limits. For this reason, the following relation for the dependence of the screening parameter r_s/D on non-ideality is recommended [GLR 83]

$$\frac{r_s}{D} = x = a \cdot \exp(-b n_D) + 1 \tag{2.80}$$

with

$$\left.\begin{aligned} a &= 5.0\,, \\ b &= 2/3 \end{aligned}\right\} \text{for} \quad n_D > 1$$

which corresponds to the solid curve in figure 2.6. In that way we have now reliable information about the macroscopic effects of charged particle interactions in a comparatively wide range of parameters. The relation (2.80) should be correct independently of the plasma composition and the gas species.

In addition to the experimental values, figure 2.6 contains the theoretical result of KAKLYUGIN and NORMAN [KN73] who made an ad hoc correction of the screening length taking for the basis

$$\frac{r_s}{D} \approx 1 + \frac{\tau_1}{\tau_2} \tag{2.81}$$

τ_1 being the time for establishment of an equilibrium DEBYE shielding cloud and τ_2 the time for establishment of an equilibrium momentum distribution. In ideal plasmas ($\mu \ll 1$) this ratio tends to unity since $\tau_1 \ll \tau_2$ because of $\tau_1 \sim \nu_p^{-1}$ and $\tau_2 \sim \nu_p^{-1}/\mu$ where ν_p is the plasma frequency. Explicitly, KAKLYUGIN and NORMAN, using for τ_1, τ_2 the expressions given in [Sp56], obtained

$$\frac{r_s}{D} = x = 1 + \frac{1}{n_D} + \frac{1}{2.5 n_D} \ln\left(n_D \frac{r_s}{D}\right). \tag{2.82}$$

As can be seen from the figure, the values of $x(n_D)$ predicted by (2.82) correspond to the general trend of the measured $x(n_D)$-function confirming the choice of the screening length as an appropriate parameter to account for nonideal behaviour.

In the method of KULIK et al. [Ku72, KKR72, KRR77] the screening length was taken to be of the form

$$r_s = \alpha(\gamma)\, n_e^{-1/3}, \tag{2.83}$$

that is, the DEBYE radius in (2.79) was simply replaced by the mean interparticle distance. Again α is a dimensionless function to be determined empirically. As a result, the following relation was found to describe the COULOMB component of the electrical conductivity [KKR72]

$$\sigma^*_{Ku} = \frac{0.487}{\ln\left(1.64\alpha(\gamma)/\gamma\right)} \tag{2.84}$$

where $\alpha = \text{const} \approx 10$ successfully averages the measurements in the range $\gamma \lessapprox 0.1$. Evidently this expression suggests non-ideality corrections which significantly exceed those reported in figure 2.6. However, it is most likely that this discrepancy is associated with uncertainties in the interpretation of a particular cesium experiment [EKKR71] which was the main justification for the model curve proposed in [KKR72].

Some other ad hoc methods to keep the transport coefficients from giving unphysical results in high-density plasmas can be found in [MKG75, Gur79]. Mitin et al. [MKG75], to reproduce their experimental results for argon in the range $n_D = 0.3 \ldots 3$, have taken the cut-off length to be of the following form

$$r_s = \begin{cases} D & \text{if } D > n_i^{-1/3} \\ n_e^{-1/3} & \text{if } D < n_e^{-1/3}, \end{cases} \tag{2.85}$$

and the Coulomb logarithm appearing in (2.76) was replaced by $\ln\sqrt{1 + \Lambda^2}$ to avoid the divergence at $\ln \Lambda \to 0$. Both corrections lead to values lower than the SPPITZER conductivity and may be interpreted to be due to an enhancement of the ion transport cross-section at increased plasma parameters.

In the treatment of GUREVICH [Gur79] a number of experimental data have been analyzed in terms of a maximum and minimum cut-off distance for the COULOMB collision cross-section. The minimum length $b_{\min}$ was taken equal to $e^2/12\pi\varepsilon_0 kT$ inspired by the widespread opinion that the argument Λ $(= 12\pi\varepsilon_0 DkT/e^2)$ under the logarithm in (2.76) would be the ratio of an upper and lower limit of integrals entering the calculation of this cross-section. Actually, $e^2/12\pi\varepsilon_0 kT$ corresponds to the mean impact parameter for a $90°$ deflection in an $e - i$ collision, and the lower limit of the binary collision cross-section (2.44) is zero. Hence, the formal attempt to explain quantitatively the measured conductivities by redefining appropriately $b_{\min}$ may lead to serious mistakes.

2.3.3. Electron-Ion Correlation Effects

In the case of weakly nonideal plasma there exist different theoretical approaches ([KN73, VK77b, Kh78]) which explain the phenomenon of the nonideal plasma conductivity on the basis of electron-ion correlation effects. The concept implies that quasi-bound particle states are in question in the process of strong interaction leading to an elimination of a fraction of electrons from the charge transport. It is well known from the literature that energy states with small binding energies may be considered to be free as well as bound which indeed makes the limit between free and bound particle states an uncertain quantity (see, e.g., [EK66a, Günd70]). Moreover, whether a particle is to be attributed to either of the two groups can depend on the effect under consideration and different values of the 'free-bound limit' may be expected for different phenomena.

Kaklyugin and Norman [KN73] estimated the effects of spatial correlation between the colliding particles taking for the basis the following expression for the level density of free electron states

$$\varrho(E) = \frac{1}{h^3} \int \mathrm{d}^3p \int \mathrm{d}u F(u)\, \delta\left(E - \frac{p^2}{2m_e} - u\right) \qquad (2.86)$$

with

$$F(u) = \left\langle \delta\left(u - \sum_i U(r_i)\right)\right\rangle .$$

E is the energy and p the momentum of the electron, $U(r_i)$ is the electrostatic interaction energy of the electron with the i-th particle, and the brackets indicate that an ensemble average over the particle coordinates has to be taken in computing $\varrho(E)$. The significance of the function $F(u)$ is that $n_e F(u)\,\mathrm{d}u$ gives the probable number of electrons which have an interaction energy in the interval between u and $u + \mathrm{d}u$. If the term $\sum U(r_i)$ is suppressed in (2.86), then we obtain

$$\varrho(E) = \frac{1}{h^3} \int \mathrm{d}^3p\, \delta\left(E - \frac{p^2}{2m_e}\right), \qquad (2.87)$$

and after doing the integration

$$\varrho(E) = \frac{2\pi}{h^3} (2m_e)^{3/2} \sqrt{E} \,. \qquad (2.88)$$

The relation (2.88) will be recognized as the level density of free electron states in the case of negligible interaction energy of the individuals. However, to obtain the correct form of $\varrho(E)$ the expression (2.88) has still to be multiplied by the internal sum of states which is $U_e = 2$ for free electrons.

In their calculations, Kaklyugin and Norman have evaluated the expression (2.86) under the following assumptions:

(i) neglect of electron-electron interactions (Lorentz plasma),

(ii) interaction between electrons and ions on the basis of a cut-off Coulomb potential terminated at $r = r_s$ where the value of r_s was determined from the relation (2.82), and

(iii) assumption of a Gaussian profile for the probability function $F(u)$.

The level density $\varrho(E)$ is related to the distribution function of electrons by

$$n_e(E) = \varrho(E) \exp\left(\frac{\mu - E}{kT}\right) \tag{2.89}$$

where μ is the chemical potential. Within the framework of the proposed model the high-density corrections to the chemical potential were estimated to be

$$\mu_0 - \mu = \frac{xkT}{6n_D} = x\,\frac{e^2}{8\pi\varepsilon_0 D} \tag{2.90}$$

where μ_0 is the chemical potential of the ideal gas and $x = r_s/D$ according to (2.82). In this way KAKLYUGIN and NORMAN derived for the distribution function of electrons the curves presented in figure 2.7. It was suggested by the authors to consider the electrons with $E < 0$ as trapped by ions resulting in their elimination from the charge transport. For this reason the following approximative formula for the electrical conductivity was recommended in [KN73]

$$\sigma = \sigma_0\,\frac{\exp\,(-r_s/6n_D)\,(1 - r_s/72Dn_D)}{1 + \ln\,(r_s/D)/\ln\,(\sqrt{2}\,9n_D)} \tag{2.91}$$

where σ_0 is the expression for the ideal conductivity of a fully ionized LORENTZ plasma with $r_s = D$. The numerator in (2.91) is due to the localization of a fraction of free electrons, whereas the denominator accounts for the enhancement of the ion transport cross-section due to the incomplete screening conditions in nonideal plasmas. Applying the relation (2.82), for $n_D = 1$ the numerical values of the respective corrections are 0.65 and 1.33.

In figure 2.8 the results of KAKLYUGIN and NORMAN are shown together with the experimental values of RADTKE and GÜNTHER [RG76] for hydrogen, and the agreement between theory and experiment seems very reasonable. However, it will be emphasized in this connection that the treatment in [KN73] requires a further consideration. The model assumes that scattering on an electron-ion pair takes place in the same manner as on a COULOMB potential. In this case the electrical conductivity can be written as

$$\sigma \sim \frac{n_e - n_{loc}}{(n_i - n_{loc})\,Q_i + n_{loc}Q_{loc}} = \frac{n_e - n_{loc}}{n_iQ_i} \tag{2.92}$$

(n_{loc} and Q_{loc} being the number density and transport cross-section of the pairs)

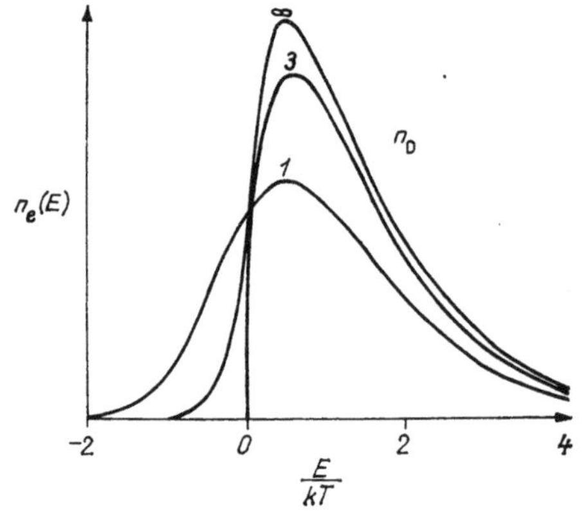

Fig. 2.7. Electron energy distribution function in weakly nonideal plasmas according to [KN73]

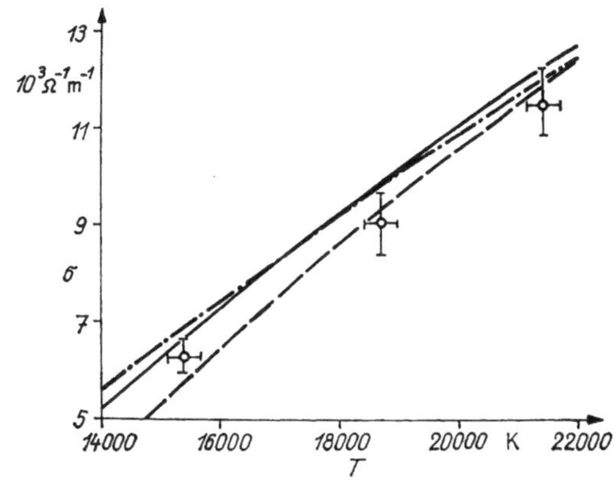

Fig. 2.8. Electrical conductivity for a hydrogen plasma with $p = 1$ MPa plotted against temperature; —·—·— calculation according to SCHIRMER and FRIEDRICH [SF58a, SF58b] including the corrections of KAKLYUGIN and NORMAN [KN73], ——— calculation according to SCHIRMER and FRIEDRICH [SF58a, SF58b] including excited atoms as well as non-ideality corrections proposed by equation (2.80), — — — according to KHOMKIN [Kh78], ⊹ experiment [GPPR76]

predicting indeed a reduction of σ by the factor $(n_e - n_{loc})/n_i$. On the other hand, if the electron-ion pair behaves like an atom with respect to a colliding particle the second term in the denominator can be neglected because $Q_i \gg Q_a$. Then we have

$$\sigma \sim \frac{1}{Q_i}$$

and non-ideality becomes apparent only due to the enhancement of the ion transport cross-section. However, it is also most likely that the pairs have cross-sections similar to those of excited atoms. As we have seen in 2.2.3.2. they are essentially greater in magnitude than the ground state cross-sections which makes the assumptions in [KN73] fairly realistic.

Plotted also in figure 2.8 is the theoretical result of KHOMKIN [Kh78] who introduced in the calculation of the electrical conductivity not only correlated pairs of negative interaction energy but also those of positive energy (collision complexes). The life time of the collision complexes cannot any longer be considered as small to neglect both the impact between them and a third particle and the elimination of a fraction of electrons from the charge transport. To calculate the number densities of correlated electron-ion pairs the concept of the density of pair energy states was applied in [Kh78, VK76, VK77a, VK77b]. By analogy with (2.86) the level density of a pair with the interaction energy E of the electron and its nearest ion can be written as

$$\varrho(E) = \frac{1}{h^3} \int \mathrm{d}^3p \int \mathrm{d}u \; F(u) \left[\delta\left(E - \frac{p^2}{2m} - u\right) - \delta\left(E - \frac{p^2}{2m}\right)\right] \tag{2.93}$$

where in this case $F(u)\,\mathrm{d}u$ is the probability that the mutual potential energy of the electron and its nearest ion is in the interval between u and $u + \mathrm{d}u$. As a function of the limits in the following expression

$$n_{1,2} = \int_{E_1}^{E_2} \varrho(E) \exp\left(\frac{\mu - E}{kT}\right) \mathrm{d}E \,, \tag{2.94}$$

VOROBEV and KHOMKIN distinguished four particle groups.

(i) States with large negative energies ($E < -kT$) were considered to be unexcited atoms the density n_a of which can be obtained by integrating from $-\infty$ to $-kT$. Unexcited atoms behave like one particle with respect to a colliding electron, and the corresponding momentum transfer cross-section may be obtained from the empirical expression (2.61).

(ii) Electron-ion pairs the energy of which is $-kT < E < -e^2/(4\pi\varepsilon_0 r_m)$ are excited atoms (r_m being the mean interparticle distance), and it has been suggested by the authors to consider electron scattering in this case as scattering on a COULOMB potential.

(iii) Quasi-bound particle states have an energy of $-e^2/(4\pi\varepsilon_0 r_m) < E < 0$ according to VOROBEV and KHOMKIN. They are characterized by the fact that the electron of the pair moves along elliptical path segments changing the nearest neighbour. In this way the trajectory is infinite even if the energy is negative.

(iv) Collision complexes finally represent correlated electron-ion pairs of positive interaction energy. The number density of this type of pairs is obtained by integrating the above expression from 0 to ∞.

Collision complexes and quasi-bound particles are unstable states, and their contribution to the electrical conductivity is a function of their lifetime. In [VK77b] the lifetime of the electron-ion pairs with $|E| \approx 0$ was estimated to be

$$\tau_0 \approx \frac{e^2}{4\pi\varepsilon_0} \cdot \frac{4\gamma^{-3/2}}{v_T\, kT} \tag{2.95}$$

where v_T is the thermal velocity. The ratio of the complex lifetime and the collision time τ_C ($\approx L/v_T = e^2/4\pi\varepsilon_0 kTv_T$) becomes therefore

$$\tau_0/\tau_C \approx 4\gamma^{-3/2} \gg 1$$

for $\gamma \leqq 0.1$. This expression shows that scattering on both quasi-bound particle states and collision complexes can be considered as scattering on one particle.

Within the framework of the proposed model the number density of free electrons contributing to the electrical conductivity becomes

$$n_e^f = n - n_a - n_a^* - n_{qb} - n_{cc} \tag{2.96}$$

where n is the density of heavy particles including free ions and all the particles that form pairs with electrons, and n_a^*, n_{qb}, and n_{cc} denote the number densities of excited atoms, quasi-bound particle states and collision complexes, respectively. It should be mentioned once more that electron scattering in the cases (ii), (iii), (iv) was considered as scattering on a COULOMB potential, and for this reason the following correction to the SPITZER conductivity (2.37) is proposed in [VK77b, Kh78]

$$\sigma = \sigma_{Sp} \frac{n_e^f}{n - n_a} q_\sigma \tag{2.97}$$

with

$$q_\sigma = \frac{1}{6(kT)^4} \int_0^\infty \frac{E^3 \exp(-E/kT)\, dE}{1 + \dfrac{\gamma_E n_a Q_a^{(1)}(E)}{(n - n_a)\, Q_i(E)}} . \tag{2.98}$$

The expression for q_σ was derived from a Frost-type interpolation formula including effects of electron-ion, electron-atom, and electron-complex collisions. γ_E is a factor expressing the effect of electron-electron collisions and is equal to 0.5816 according to [SH52]. If the effect of neutral particles is neglected in the above expression ($n_a = 0$), then q_σ approaches unity, and the remaining correction term to the SPITZER relationship is exclusively associated with the specific features of the electron interaction with the scatter pairs. VOROBEV and KHOMKIN, using the concept of the density of pair energy states, obtained for it

$$\frac{n_e^f}{n - n_a} = 1 - 0.47\mu^{2/3} - 0.33\mu^{4/3} \tag{2.99}$$

where μ is the plasma parameter (1.6).

It should be emphasized here that the relation (2.99) predicts corrections which are smaller in magnitude than those reported in [KN73]. None the less, for the example dealt with in figure 2.8 we find that the data of VOROBEV and KHOMKIN are somewhat lower in value than those predicted when applying the correction (2.91). The reason is that the calculations in [Kh78] have been carried out with coefficients c, d, f for the ground state momentum transfer cross-section differing from those listed in table 2.2. This can lead to a substantial variation of the electrical conductivity since the atomic contribution is relatively large for the parameter range under consideration. Moreover, it is to be mentioned that interpolation formulas such as (2.98) cannot naturally provide high accuracy if the various collision processes make quantitatively comparable effects on the electron transport.

Quasi-bound particle states and collision complexes are characterized by the fact that the electron moves along elliptic or hyperbolic path segments changing the nearest neighbour. To estimate the contribution to the conductivity made by these two categories of electron motion the following approximative formula was applied in [VK77b, Kh78]

$$\Delta\sigma = \frac{e^2(n_{qb} + n_{cc})}{m_e \nu} . \tag{2.100}$$

The collision frequency was assumed to be $\nu \approx \tau_0^{-1}$ where τ_0 is the complex lifetime according to (2.95), and for the calculation of the particle number densities in (2.100) again the concept of the density of pair energy states was applied. In this way the authors arrived at

$$\frac{\Delta\sigma}{\sigma_{Sp}} = \frac{0.11\mu^{5/3} + 0.08\mu^{7/3}}{\ln(3/\mu)} . \tag{2.101}$$

This expression shows that for $\mu < 1$ the contribution of quasi-bound particle states and collision complexes to the charge transport can be neglected.

Calculations of the electrical conductivity have been carried out by VOROBEV and KHOMKIN for hydrogen, argon, krypton and xenon, and in all cases an excellent agreement with experimental results was noticed. Nevertheless, the proposed model is still a crude approximation to the true behaviour and the introduction of the scatter pairs remains necessarily artificial and afflicted with inherent uncertainty.

Actually, VOROBEV and KHOMKIN distinguished two groups of electron-ion pairs only: strongly interacting particle states ($E < -kT$) which behave like atoms with respect to a colliding electron on the one hand, and weakly interacting particle states ($E > -kT$) which behave like COULOMB scatters on the other. From the physical point of view the COULOMB cross-section represents an upper limit for electron momentum transfer in plasma, and it is therefore most likely that the effect of the pairs listed under (ii), (iii), (iv) is overestimated. On the other hand, in describing the process of momentum transfer with ions the authors assuming ideal plasma conditions have terminated the interaction at the DEBYE length. This approach, however, is not immediately applicable to the plasmas of current interest, since the shielding equilibrium characteristic of nearly ideal plasmas will hardly be expected at high densities.

Finally, we remark that the complex number densities themselves cannot be considered consistently defined in the case of a plasma since the expression (2.93) uses a very simplified model neglecting the interaction of the pairs with the surrounding charge carriers.

2.3.4. Corrections for Stronger Non-Ideality

The aim of this chapter is to give a brief presentation of high-density corrections relevant to strongly nonideal plasmas. For a detailed survey which includes all these effects, the reader is referred to [EFKK84].

A comparison between experimental values and different theoretical approaches developed to the plasma conductivity shows that the measurements in the weakly nonideal region can satisfactorily be described within the framework of the binary collision model taking into account the non-ideality corrections discussed in sections 2.3.2. and 2.3.3. Nevertheless, there are additional effects which are expected to alter the electric properties of real plasmas particularly in the region of stronger non-ideality. Among them the numerically most important effects are related to the dynamical screening and the DEBYE-ONSAGER relaxation force. The latter effect was first studied in investigating the properties of ionic solutions and is related to the relaxation of the (DEBYE) shielding cloud in the presence of an external electric field. The dynamical screening is due to plasma microfield distortions by the colliding particle resulting in a dynamically instead of a statically screened potential. Both effects have been studied classically as well as quantum-mechanically in a series of papers (see, e.g., [ER79, REK80] and references therein).

The calculations by EBELING and RÖPKE [ER79] are carried out using the kinetic equations for the distribution function after KLIMONTOVICH [Kli75] as well as the correlation function theory. The KLIMONTOVICH equation incorporates the effect of the external field and treats the collective shielding effects by a collision integral of the LENARD-BALESCU form. Its essential limitation is the restriction to weak deflections, and therefore an appropriate cut-off r_{min} must be applied to exclude the contributions of collisions with small impact parameters. To obtain a solution to the kinetic equation the standard approaches were applied in [ER79], and in this way the following expression for the conductivity of a completely ionized plasma was derived

$$\sigma = \frac{1}{f} \cdot \frac{(4\pi\varepsilon_0)^2 \, (kT)^{3/2}}{e^2 m_e^{1/2}} \cdot \frac{1 - Q}{L} \,. \tag{2.102}$$

L is the generalized COULOMB logarithm and Q the relaxation function, so called because this term is due to the effect of the external field. For the numerical factor f the usual CHAPMAN-ENSKOG scheme yields $f = 1.69$. Within a simple approximation the functions L and Q were found as

$$L = \int_0^{k_m} \frac{k^3 \left(\frac{V(k)}{4\pi}\right)^2 \mathrm{d}k}{1 + \frac{V(k)}{4\pi D^2}} \tag{2.103}$$

and

$$Q = \frac{e^2}{12(2\pi)^4 \varepsilon_0 D^2 kT} \int_0^{\infty} \frac{k^2 V^2(k)\, \mathrm{d}k}{1 + \frac{V(k)}{4\pi D^2}} \tag{2.104}$$

respectively, where $V(k)$ is the FOURIER transform of the potential $V(r)$, k_m a cut-off value to ensure the convergence of the integral (2.103), and D the DEBYE length. For the special case of a COULOMB interaction potential the FOURIER transform is given by $V(k) = 4\pi/k^2$, and when this relation is applied then the above expressions read as

$$L = \ln \sqrt{1 + k_m^2 D^2} \tag{2.105}$$

and

$$Q = \frac{e_2}{24\pi\varepsilon_0 DkT} = \frac{\mu}{6} \tag{2.106}$$

respectively, where μ is the plasma parameter (1.6). It was suggested in [ER79] to take the cut-off value k_m equal to $12\pi\varepsilon_0 kT/e^2$ since in this case the generalized COULOMB logarithm approaches the SPITZER expression in the ideal limit. It turns out that both these effects have obvious implications only in the strongly nonideal region. In weakly nonideal plasmas the relative deviation of (2.105) as compared with $\ln \Lambda$ corresponding to the statically screened Coulomb potential is small and does not exceed 1%. We can assume therefore that the statically screened potential gives the full description of the plasma in this case. The DEBYE-ONSAGER relaxation force gives rise to the factor $1 - Q$ in the expression (2.102) which decreases the conductivity by about 5% at $n_D = 1$.

Quantum-mechanical expressions for the generalized COULOMB logarithm L and the relaxation function Q have been derived in [ER79, REK80] using the SUBAREV formalism [Sub76] which connects the electrical conductivity with correlation functions within the frame of a linear response theory. The basic equation for the conductivity of an electron system embedded in scatters is given by

$$\sigma^* = \frac{1}{f} \cdot \frac{4\sqrt{2\pi} N_e^2 e^4 m_e^{1/2}}{\Omega (kT)^{3/2}} \cdot \frac{1 + (3N_e m_e)^{-1} \langle \boldsymbol{P}; \boldsymbol{F} \rangle}{\langle \boldsymbol{F}; \boldsymbol{F} \rangle} \tag{2.107}$$

where Ω is the system volume, N_e the total number of electrons, $\boldsymbol{P}$ the total impulse of the electron system, and $\boldsymbol{F}$ the total force acting on the electrons. The correlation functions $\langle \boldsymbol{F}; \boldsymbol{F} \rangle$ and $\langle \boldsymbol{P}; \boldsymbol{F} \rangle$ have to be derived from an expression of the following form

$$\langle A; B \rangle = \int_0^{\infty} \mathrm{d}\tau\, \mathrm{e}^{-\varepsilon\tau} \int_0^{kT} \mathrm{d}\lambda\, tr \left\{ \frac{1}{Z} \mathrm{e}^{-\frac{H}{kT}} A(-\tau - \mathrm{i}\hbar\lambda)\, B \right\} \tag{2.108}$$

where H is the system Hamiltonian, Z the partition function, and $\varepsilon > 0$ a small convergence parameter. In [ER79] the correlation functions occurring in (2.107) were evaluated within the frame of a perturbation theory, and to account for screening effects the COULOMB interaction potential was replaced by a statically screened one. In this way quantum-mechanical expressions were obtained which decrease L and Q in comparison to their classical values

$$L = \int\limits_0^{k_m} \frac{k'^3 \, \mathrm{e}^{-\frac{\hbar^2 k'^2}{8 m_e k T}} \left(\frac{V(k')}{4\pi} \right)^2 \mathrm{d}k'}{1 + \frac{V(k')}{4\pi D^2}} \tag{2.109}$$

$$Q = \frac{e^2 kT}{12(2\pi)^4 \, \varepsilon_0 \hbar m_e D^2} \int\limits_0^\infty \tau^2 \, \mathrm{d}\tau \int\limits_0^\infty \frac{k'^4 V^2(k') \, \mathrm{d}k'}{1 + \frac{V(k')}{4\pi D^2}} \mathrm{e}^{-\frac{k'^2 \tau^2 kT}{2 m_e}} \sin \left(\hbar k'^2 \tau / 2 m_e \right) . \tag{2.110}$$

However, in the range $\mu < 1$ the effect of quantum corrections is small, and both expressions coincide within a few per cent with the respective classical limit results (2.105) and (2.106).

An improved calculation was done by RÖPKE et al. [REK80] who seperated the force-force correlation function into three independent parts corresponding to a seperation of the particle interaction into an electron-ion, electron-electron, and free electron contribution, respectively. In their calculations, the authors, neglecting screening effects in the treatment of the electron-ion interaction, obtained for the COULOMB logarithm L an expression which corresponds to a LENARD-BALESCU type expression for the collision term. The numerical results show the particular feature that the reduced electrical conductivity σ^* is basically a function of both the plasma parameter μ and the temperature. Similar results were obtained by RÖPKE et al. [RMKK79] and more recently by MEISTER and RÖPKE [MR82]. In the latter paper the force-force correlation function was evaluated in the BORN approximation, whereas in [RMKK79] the conductivity was derived within simple approximations from the long wave limit of the complex dielectric function. However, it should be mentioned here that a spreading-out for σ^*-curves at different tempeartures occurs also within the kinetic approaches if, for instance, the cross-section for a screened COULOMB potential in first BORN approximation is applied to the calculation of electron transport coefficients (see, e.g., [EK79]). The sole dependence of the reduced conductivity on the plasma parameter μ may therefore be understood as a property of the COULOMB potential connected with a DEBYE cut-off (see eq. (2.35)), and gradual variations in the behaviour are expected in going from Coulombic to non-Coulombic potentials.

The theories mentioned in this chapter will attain essential importance for the quantitative understanding and control of the transport phenomena in nonideal plasmas when they succeed in including the interaction of the charge carriers with neutrals and excited particles, and when the many body interactions are comprehended such that artificial assumptions about the acting potentials or cut-off procedures become dispensable.

3. Production of Weakly Nonideal Plasmas in the Laboratory

3.1. Behaviour of Nonideal Plasmas under Experimental Conditions

Ideal plasmas of laboratory dimensions are usually weakly ionized and, with respect to the main part of radiated power, optically thin. The power necessary to maintain such a discharge is small enough for stationary operation and, in the case of electrical discharges, the radial temperature profile is determined essentially by heat conduction and diffusion processes. On the contrary, simple estimates and more detailed calculations for many nonideal plasmas show that losses by heat conduction and diffusion may be neglected compared to radiation losses and that radiation cooling normally exceeds by far the level attainable in the stationary mode.

3.1.1. Loss Mechanisms

We will estimate the contribution of the heat conduction to the energy balance for a nonideal plasma and, for simplicity, restrict this consideration to a cylindrical electrical discharge. The power balance of such discharges is given by the following equation

$$N_{el} - \operatorname{div} F_C - \operatorname{div} F_R = 0 \tag{3.1}$$

where N_{el} is the electric power per unit volume, and F_C and F_R are the fluxes of heat and radiation, respectively. For cylindrical symmetry we get the well-known ELENBAS-HELLER equation

$$\sigma E^2 + \left[\lambda \cdot \frac{\partial^2 T}{\partial r^2} + \frac{\partial \lambda}{\partial r} \cdot \frac{\partial T}{\partial r} + \frac{\lambda}{r} \cdot \frac{\partial T}{\partial r}\right] - S = 0 \tag{3.2}$$

Here σ, λ, and r are the electrical and thermal conductivity, and the radial coordinate, respectively. S is the radiation flux

$$S(T) = \operatorname{div} F_R\,. \tag{3.3}$$

To get an impression of the contribution of the different mechanisms to the balance equation (3.1) we simplify the problem as follows:

(i) The degree of ionization is sufficiently high to assume that both heat conduction and charge transport are maintained by the electrons alone.

(ii) To calculate the derivatives of $T(r)$ in (3.2) we assume a quadratic decay of T:

$$T = T_0(1 - r^2/R^2) \tag{3.4}$$

where R is the radius of the discharge column.

(iii) We suppress the logarithmic dependence of σ on T and write

$$\sigma = \sigma_0 T^{3/2}\,. \tag{3.5}$$

We use the WIEDEMANN-FRANZ law adapted for plasma kinetics as proposed by SCHIRMER [Schi60] to obtain a relation between thermal and electrical conductivity

$$\frac{\lambda}{\sigma} = \frac{\Lambda_\lambda}{\Lambda_\sigma} \cdot 2 \left(\frac{k}{e}\right)^2 \cdot T \tag{3.6}$$

where Λ_λ and Λ_σ are the mean free path lengths for the transport of heat and charge. ENSKOG [En12] found for the quotient

$$\frac{\Lambda_\lambda}{\Lambda_\sigma} = \frac{\nu}{\nu - 1}$$

where ν is the power of the interaction law. For COULOMB forces we get with $\nu = 2$

$$\frac{\lambda}{\sigma} = 4 \left(\frac{k}{e}\right)^2 \cdot T . \tag{3.7}$$

We eliminate λ according to (3.7) and express the derivatives in (3.2) using (3.4) and finally find for (3.2)

$$\sigma \left\{ E^2 - 2 \left(\frac{k}{e}\right)^2 \cdot \frac{T_0^2}{R^2} \right\} - S = 0 . \tag{3.8}$$

Comparing the terms in braces we find for typical values $T \approx 10^4$ K, $E = 10^4$ V/m, and $R = 10$ mm

$$\frac{E^2}{2 \left(\frac{k}{e}\right)^2 \cdot \frac{T_0^2}{R^2}} \approx 10^4 . \tag{3.9}$$

Consequently, in typical nonideal plasmas the heat conduction can be neglected compared to the radiation losses. In agreement with experience, this statement becomes incorrect if the temperature $T > 10^5$ K and $R < 1$ mm (hot sparks) or if the field strength decreases to $E \approx 10^3$ V/m and the degree of ionization becomes lower. Then assumption (i) and (3.6) fail and we get the stationary arc conditions with dominant heat conduction.

It should be mentioned that this treatment is not consistent because the temperature profile $T(r)$ should be a result of the calculation rather than one of the preconditions. However, the resulting temperature profile is smoother than assumed in (3.4) because of the small heat conduction, so the gradients become smaller and the quotient of the electric power and the heat losses is even larger than in (3.9).

3.1.2. Radial Temperature Profiles

The radial profiles of the plasma temperature and, consequently, of the electrical conductivity and the current density are of basic importance for the calculation of transport properties from experimental data. The well established stationary arc plasma physics has developed methods for obtaining temperature profiles from spectroscopic end-on and side-on observations [Loc68, Ne75]. The usual narrow bell shape of these profiles which is determined by the relation between heat conduction and electrical conductivity requires a high experimental spatial resolution and limits the attainable accuracy of the derived transport properties. As we have shown in section 3.1.1., above a certain value of the electrical field strength (or also of the electrical power) heat conduction processes do not influence the energy balance of the discharge and (3.2) reads as

$$\sigma E^2 - S = 0\,. \tag{3.10}$$

The radiation term can be expressed by the emission coefficient $\varepsilon(\lambda, T)$, the absorption coefficient $\varkappa(\lambda, T)$, and the PLANCK function $B(\lambda, T)$

$$S = \int_0^\infty \int_0^{4\pi} \frac{1}{\varkappa(\lambda, T)} [B(\lambda, T) - I(\lambda, \omega)]\, d\omega\, d\lambda \tag{3.11}$$

where ω means the solid angle.

S can be considered to be a unique function of temperature if for the radiation intensity $I(\lambda, \omega)$ the following relation holds

$$I(\lambda, \omega) \ll B(\lambda, T)\,, \tag{3.12}$$

i.e. the radiation in the plasma is essentially optically thin in all directions. Under such conditions (3.10) contains only quantities which depend on the plasma temperature at the considered position rather than on the spatial temperature distribution too. Thus, all volume elements of the plasma are in the same situation concerning production and loss of energy; there is no reason for any gradients of the temperature except the extreme boundary layers. Experimental investigations [Gü70b] really show that such profiles are extremely flat (curve 4 in figure 3.1) as far as (3.12) is valid. The characteristic curvature of the temperature profiles for the lighter rare gases and for hydrogen (curves 2, 3 in figure 3.1) is no contradiction and may be

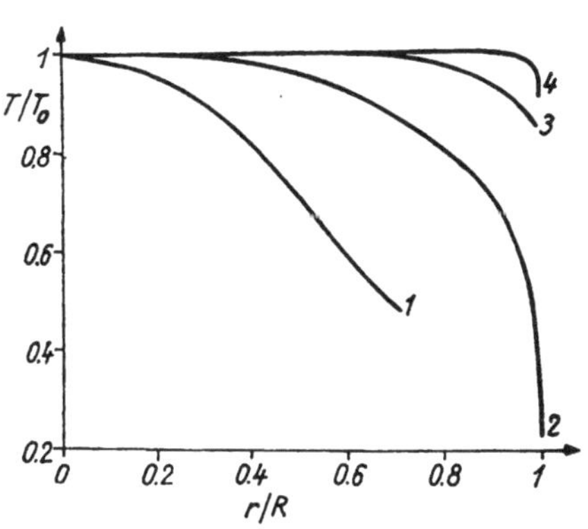

Fig. 3.1. Radial temperature profiles of cylindrical stationary and pulsed arcs; 1 stationary argon cascade arc: $p = 0.1$ MPa, $T_0 = 26\,800$ K [St69], 2 pulsed hydrogen arc: $p = 1$ MPa, $T_0 = 21\,500$K [RG76], 3 pulsed argon arc: $p = 2$ MPa, $T_0 = 15\,920$K [GLR83], 4 pulsed xenon arc: $p = 1$ MPa, $T_0 = 11\,200$ K [Gü70b]

explained from the different radiation characteristics of these plasmas: with increasing atomic number the fraction of the total emitted power which originates from continuous transitions (bremsstrahlung and recombination continuum) becomes larger. These emission processes are more or less equally distributed over the whole spectrum and run under optically thin conditions up to rather high power levels [GLR83, Gü73]. The result is a volume specific energy balance which is independent of the radius and yields very flat radial temperature profiles. In plasmas of the lighter rare gases or of hydrogen the continuous radiation is of less importance, thus, in discharges of the same power dissipation more energy is emitted by strong lines which are optically thick. In such plasmas (3.12) is violated, and the energy balance is strongly influenced by radiation diffusion effects which produce the curvatured profiles 2, 3 shown in figure 3.1. According to this explanation the curvature increases for decreasing atomic weights and is most apparent for hydrogen. It should be mentioned that above a certain level of the specific electric power even xenon plasmas should assume a curvatured profile again when the discharge column becomes optically thick for the main part of the emitted radiation [Lit73]. The temperature profiles of the pulsed argon arc investigated in [GLR83] exhibit all types of energy balance discussed here for one gas at different power levels: the lowest curve represents the influence of heat conduction, at higher levels the dominating optically thin radiation produces a flat profile, and the upper profile is curvatured due to the radiation diffusion.

3.1.3. Time Development and Stability

The accuracy and reliability of experimental results depend essentially on the behaviour of the discharge in the rise and decay period in case of pulsed arcs and on the conditions to reach a stationary period. For dc arcs there is a lot of practical experience [Loc68, Ne75] and theoretical considerations [Low70, Low74, RSL78] about the region of existence. The lower limit of plasma temperature at a given pressure is indicated by irregularities and by the tendency of the arc to extinguish due to strong constrictions of the arc column and the connected high energy losses by heat conduction. The maximum temperature is given by the limited heat transfer from the discharge channel, by the tendency to form arcs between the channel plates at high electric field strength, and by instabilities of the arc column due to magnetic forces which can be suppressed to some extent by an optimized gas flow in the discharge channel [BS72].

In pulsed arcs the regions of existence are determined by completely different aspects which are discussed in detail in section 3.2.5. Generally, they are extended towards higher plasma temperatures and higher power dissipation due to the isolation of the plasma from the discharge vessel by a thin cold gas layer. This isolation is effective until the gas layer is destroyed by heat conduction after some milliseconds; therefore, the thermal load of the tube walls in discharges of millisecond duration is determined mainly by the absorbed radiation rather than heat conduction. Thus, the power limit of a pulsed arc exceeds that of a stationary discharge by more than one order of magnitude. However, this limitation in time for powerful pulsed discharges raises the necessity to investigate the processes which lead to quasi-stationary conditions where all electric and thermodynamic parameters are in equilibrium except the heat transfer to the tube walls.

For the calculation of the time development we must modify (3.10) by adding the change of the internal energy $\mathrm{d}u(T, p)/\mathrm{d}t$ during the heating or decay process:

$$\sigma E^2 - S - \frac{\mathrm{d}u}{\mathrm{d}t} = 0 . \tag{3.13}$$

The internal energy is given by

$$u = \tfrac{3}{2}\, kT\, (n_e + \Sigma n_j) + \Sigma\, n_{j+1} \chi_j \tag{3.14}$$

where the sum is taken over the particle number densities n_i of all species with the ionization energies χ_i. Some typical values for the terms of (3.13) may clarify the situation: A xenon plasma of 1 MPa and 15000 K has an internal energy of 9 MJ/m³, the radiation losses at the same conditions are $S \approx 0.5$ TW/m³ [Gü73, KKM75] while the specific electric powers in pulsed discharges are between 10^{-2} and 10^2 TW/m³. Thus, we may distinguish between two extreme cases:

(i) The electric power is large compared with the radiation power:

$$S \ll \sigma E^2 . \tag{3.15}$$

Such discharges cannot dissipate the supplied power, they heat up very fast to rather high temperatures. The mechanism can be considered to be an adiabatic process; every discharge runs through such a state after ignition, but we refer to publications [GH75, HG77] for further information because these plasmas are completely ideal due to their high temperature.

(ii) The radiation power and the electric power are of the same order of magnitude:

$$S \approx \sigma E^2 . \tag{3.16}$$

The last expression characterizes the quasi-stationary radiation dominated plasma whose energy balance is nearly in equilibrium. To understand the most important features of such discharges we will modify the calculations of Lowke [Low70] which refer originally to the stability and radial distribution of stationary radiation dominated arcs. For the heating period we get

$$0 < \frac{\mathrm{d}u}{\mathrm{d}t} \ll S\,,\ \sigma E^2$$

and from (3.13)

$$E^2 \geqq \frac{S}{\sigma} = f(T) \tag{3.17}$$

where the right-hand expression can be understood to be a function of the temperature for constant pressure or constant particle number density. It can be seen directly that there are the following possibilities depending on the structure of $f(T)$:

(i) $\mathrm{d}f/\mathrm{d}t > 0$: there exists one stable solution which is given for the equality in (3.17), i.e. for a given field strength E (3.17) yields an equilibrium temperature, and all properties of the discharge as current density, radiation, electrical conductivity may be calculated.

(ii) $\mathrm{d}f/\mathrm{d}t < 0$: A stationary condition is not possible, the temperature increases until regions with a positive slope of $f(T)$ are reached or the field strength decreases because the external source breaks down.

(iii) $f(T)$ has a maximum: there are two solutions for T where the upper one is unstable. For increasing temperature the difference of electrical energy and radiation which is transformed into the internal energy becomes larger; thus, the heating process accelerates until $\mathrm{d}f/\mathrm{d}t$ becomes positive again.

The preferred state for the use of such discharges for technical applications as well as for basic research is the first one. It is the normal situation for thermal plasmas with a temperature in the region of $T \approx 0.1\chi_i/k$ where χ_i is the ionization energy. However, a simple estimate shows that there are regions where the second situation is realized: we equate the radiation term S in (3.17) with the recombination and bremsstrahlung continuum and write

$$S = 4\pi \int_0^\infty \varepsilon \, d\lambda \sim n_e T^{1/2} . \tag{3.18}$$

For σ we use (3.5) and find

$$f(T) \sim \frac{n_e^2}{T} . \tag{3.19}$$

In the neighbourhood of the temperature corresponding to full ionization n_e remains constant or is slightly falling with increasing temperature. Thus, $f(T)$ is falling until the ionization process of the next stage results in a positive slope of $f(T)$ again [LLF76]. The experimental result of such a situation is an increasing plasma temperature even when all external parameters are kept constant. Figure 3.2 shows this behaviour

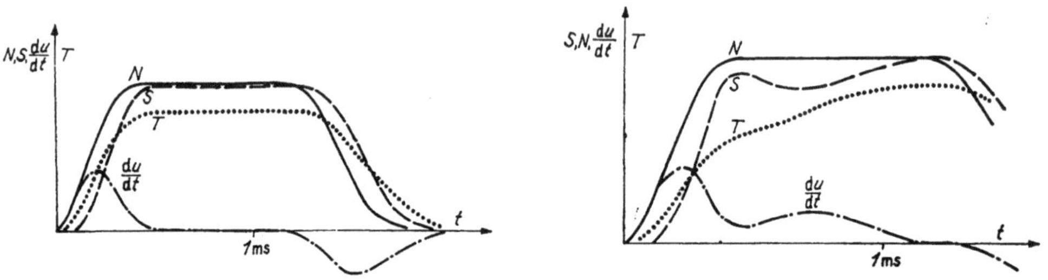

Fig. 3.2. Development in time of the electric power N, the specific radiation power S, the derivative of the internal energy du/dt and the plasma temperature T for a) a stable discharge with $df/dt > 0$ and b) an unstable dischage with $df/dt < 0$ (explanation see text)

schematically for a rectangular power pulse in the case of a stable and an unstable discharge. Obviously such discharges are not qualified for precise investigation of transport phenomena in nonideal plasmas.

Another complication is connected with the interaction between the plasma and the discharge circuit. To reach high electric powers it is necessary to match the resistance of the discharge to the impedance of the network resulting in a strong feedback of the discharge plasma to the circuit. Thus, it is not clear from the first whether a discharge under the same conditions tends to expand its radius with decreasing plasma temperature or not. It can be shown that many such conditions can be realized without violation of the energy balance. On the other hand, there are experimental indications that short-arc pulsed discharges constrict their radius very sharply without solid walls and that wall-stabilized arcs with long electrode distances tend to fill the discharge tube exactly. However, under certain circumstances such wall-stabilized arcs show irregular constrictions which are instationary in time [VSBP79]. There are indications that these irregularities are connected with shock waves [KF76a] or pressure oscillations [EK66b]. So far, their origin is not yet completely understood they should be avoided carefully in model plasmas for experimental conductivity measurements.

Preliminary estimates which include the investigation of the energy balance in radiation dominated discharges under optically thin and optically thick conditions as well as stability considerations yield the following conclusions:

Optically thin radiation dominated arcs with temperatures below $0.1\chi_i/k$ expand and decrease their temperature until they are limited by solid walls. Experimental confirmations of this theoretical result can be found on the streak photographs of GÜNTHER [Gü68] and VASILEVA et al. [VGP71]. Above the maximum temperature of the function $f(T)$ they become unstable: the plasma temperature increases without any external cause until df/dt is positive again and the introduced power is balanced again by radiation cooling.

Another situation for instability may arise in partially optically thick plasmas which are radiation dominated. It was found that in such discharges for some operating conditions there are multiple solutions for the cross-section of the discharge channel and the plasma temperature. This ambiguity is the more dominant the larger the fraction of optically thick radiation of the total emision is. In fact, we find more indications of this type of instability in discharges of lighter gases as argon than in xenon [VSBP79], and hydrogen plasmas proved to be still more unstable. According to the explanation of the stronger curvature of the radial profiles of discharges of lighter gases in chapter 3.1.2. we understand this tendency to be the result of the larger fraction of optically thick radiation at comparable total radiation powers.

3.2. Devices

Interest in non-ideality effects stems largely from the many plasma sources in which the densities reach the boundaris of several basic validity criteria in so far as theoretical treatment of the plasma is considered. Examples are plotted in figure 1.2 in terms of the range of electron concentration and temperature typical of these sources. In addition, high-density plasmas occur in a series of laboratory plasma sources such as z-pinch discharges [MN77, BB79] ($T = 10^4 \dots 10^5$ K, $n_e = 10^{23} \dots 10^{25}$ m^{-3}), laser sparks [Co72] ($T = 10^5 \dots 10^6$ K, $n_e = 10^{24} \dots 10^{27}$ m^{-3}), vacuum sparks [Co72] ($T \approx 10^7$ K, $n_e = 10^{26} \dots 10^{27}$ m^{-3}) or dense -plasma focus devices [Co72] ($T \approx 10^7$ K, $n_e \approx 10^{25}$ m^{-3}) as well as in a new generation of plasma sources which are of topical interest in fusion research (laser-compressed plasmas [At78] at $T = 10^6 \dots 10^7$ K, $n_e = 10^{26} \dots 10^{30}$ m^{-3}). Another example are the plasmas formed above the crater of vacuum arc discharges which exhibit average electron densities and temperatures of 10^{24} m^{-3} and 2 eV, respectively [PP79].

The common feature of all these plasmas is the high particle number density. Thus, more or less strong nonideal conditions determine their temporal development, the stability ranges, and their energy balance on the whole or at least in essential phases, and the transport properties considered in this review should be useful to understand and predict essential properties. A second characteristic feature of these plasma configurations is the extremely large gradient of density and temperature with respect to time and space. In this way, the description of such a plasma requires much additional information and corresponding experimental effort without the chance to reach a precision comparable to that of stationary or quasistationary plasmas. For this reason, such configurations are not at all qualified to study nonideal transport properties and were excluded from the following discussion.

The aim of this chapter is to point out a few typical techniques with respect to the generation of high-density low-temperature plasmas, as well as to offer some practical hints for handling these sources. Thereby, much preference is given to the pulsed wall-stabilized discharges which have proved extremely profitable for measuring high-temperature charge transport properties under weakly nonideal conditions.

3.2.1. Stationary High-Pressure Arcs

Stationary wall-stabilized cascade arcs proved to be very suitable for investigations on nonideal plasmas when operated at high pressures [Bau76, BDM73, BM71, BS72, GNPP78] (see also figure 3.4). Even for low ionization degrees they reach charge carrier densities of $5 \cdot 10^{23}\ \mathrm{m}^{-3}$ and γ-values of 0.12, the upper limit of these parameters being limited by the maximum thermal loss power capacity of the cascaded discharge channel of about 2 MW per m arc length and by the maximum axial field strengths at which parasitical arcs between the metal discs can be prevented (see, e.g., [MS67]). The lower limit of operation is given by the deviations from LTE which become apparent when the electron density is too small to thermalize all plasma components and excitation populations by means of interaction processes. Thus, stationary high-pressure arcs overlap the regions of the ideal plasma to beginning non-ideality; and they represent the reference condition for ideal behaviour as well as the beginning deviations from it. The identical plasma source and diagnostic tools for the whole region allow to detect even small effects with high accuracy. Such discharges may be operated in the dc mode, and all signals are observed using the well established stationary detection methods for ideal plasmas with high accuracy. The arcs have nearly cylindrical symmetry and may be investigated spectroscopically side-on using optical windows or end-on through a cold buffer gas. The tendency to produce corcscrew instabilities at higher gas pressures induced by selfmagnetic fields may be suppressed by carefully optimized gas flows in the discharge channel [BS72]. Other deviations from the cylindrical symmetry, namely the convex shape of the arc between the cascade plates, are minimized by insulating spacers; the problem remains how to get reliable radial profiles by means of side-on measurements.

The electrical field strengths necessary for the calculation of the electrical conductivity are derived from measurements of the cascade disc potentials; in this way the problem to subtract the electrode drop voltages which are normally very difficult to determine is eliminated principally. However, the finite insulation of the cascade plates due to the cooling liquid introduces uncertainties concerning the measured arc potentials which can reach some per cent. Nevertheless, the electrical conductivities derived from electrical measurements of cascade arcs are reliable within 10 ... 15% [Pl69]. The peculiarities of the experimental techniques are described in detail in several publications [BPR67, Em67, He68a, HM67, Ko71a, Ko71b, Pl69], and it is not necessary to repeat them here. Besides it should be mentioned that the plasmas produced in dc arcs are dominated by electron-neutral interaction, effects of nonideal charge carrier interaction are superimposed by the interaction between charges and neutrals because of the low degree of ionization.

3.2.2. Radiation-Generated Plasmas

Plasmas which are generated by absorption of intense radiation have been investigated carefully by GENERALOV, KOZLOV and co-workers [GZKM70, KKM74b, KKM75] and by the group of UHLENBUSCH [HMU72]. The reported parameters for pressure and temperature indicate that such plasmas are nonideal in many cases. It is very advantageous that they can be operated stationarily with cylindrical symmetry and that they are completely isolated from material walls. Thus, they allow a high purity of the plasma and a very precise spectroscopic investigation. The pronounced radial distribution of the temperature requires a good spatial resolution including Abel conversions. Measurements of the electrical conductivity have not been performed yet, they should be done contactless on the basis of inductive methods.

3.2.3. Shock Waves

Shock waves have been used successfully to produce and investigate nonideal plasmas (see, e.g., [BGLS77, GIF77, KR74, LF72, LFS70]). The temperature-pressure regions include essentially those of pulsed arc plasmas and can be selected by a proper choice of the initial conditions. An important advantage of these plasmas is their homogeneity and the possibility to determine the temperature from the shock velocity. However, there are doubts that the assumption about the conservation of energy in the shock front which is one of the basic preconditions for the derivation of the RANKINE-HUGONIOT equations can be maintained in the case of high plasma densities with corresponding high radiation losses [KR74]. Except that, it is not easy to get reliable measurements of the electrical conductivity under such conditions. The only attempt to perform quantitative conductivity measurements was reported by IVANOV et al. [IMFD76]. The shock wave produced by an explosion touches two electrodes which introduce a current into the plasma and two probes for measuring the potential

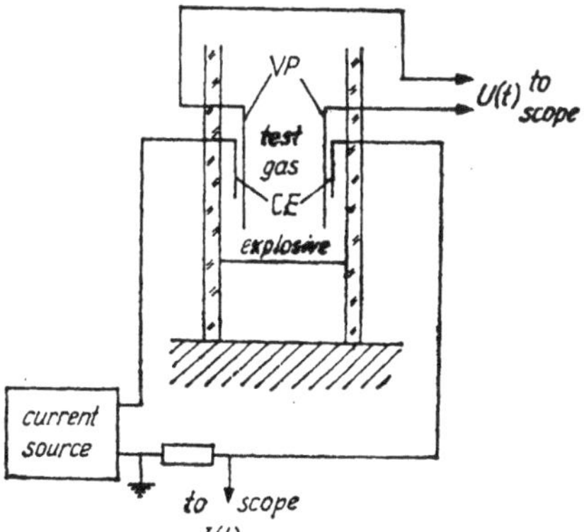

Fig. 3.3. Experimental arrangement for the determination of the electrical conductivity in shock tubes [IMFD76]; VP voltage probes, CE current electrodes

of the plasma. The geometrical cross-section of the conducting medium which is necessary for the determination of specific data as the electrical conductivity was determined by stationary measurements in electrolytes and by X-ray investigations of the ionized shock front [BGDF75]. Such experiments are of much interest because of the large range of non-ideality realized, but there are no estimates about the reliability of the derived data (see figure 3.4).

3.2.4. High-Density Plasmas Produced by Adiabatic Compression or Wire Explosion

The fast compression of a gas contained in a cylindrical steel tube by means of a pneumatic accelerated piston produces homogeneous plasmas of high pressures over a large temperature range [DDES69, DHTW77, KK74a]. The diagnostics is to some extent similar to that of pulsed electric arcs (see section 3.2.5.) apart from particular problems due to the extremely high densities and short optical path lengths. Regardless of their homogeneity and the attractive thermodynamic conditions these plasmas have not yet been used for the investigation of nonideal transport properties. Nonideal conditions are maintained also in fast discharges in the vapour of exploding wires (see, e.g., [DRRZ77, SRG77, SRGC70]). However, such plasmas are very inhomogeneous in space and instationary; therefore, reliable conductivity measurements seem to be extremely difficult. Similar conditions are present in the miscellaneous experiments described in [DRRZ77, KKPR77].

3.2.5. Pulsed Wall-Stabilized Discharges at High Pressure

Pulsed wall-stabilized discharges represent a consequent extension of the stationary plasmas to higher non-ideality parameters. They allow a considerably wider range of

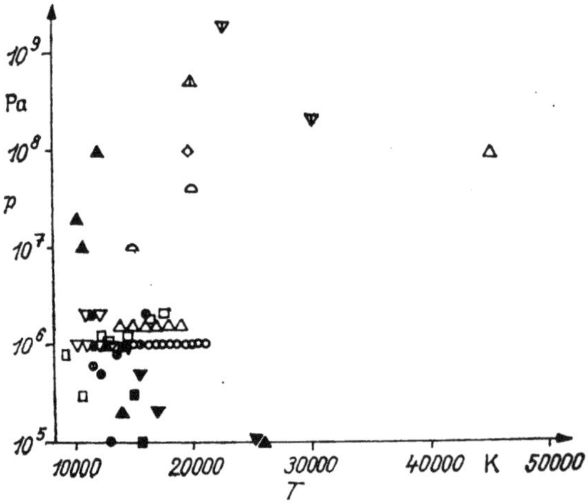

Fig. 3.4. Pressure-temperature diagram of plasmas for the investigation of transport properties;
□ Ar, Bakeev and Rovinski [BR70b], ▽ Xe, Günther [Gü70a], ■ Ar, Batenin and Minaev [BM71], ▼ Ar, Kopainsky [Ko71a, Ko71b], ▲ Ar, Bauder and Stephens [BS72], Bauder et al. [BDM73], △ Ar, Radtke and Günther [RG73], ▽ Xe, Radtke and Günther [RG73], ◡ Xe, Popović et al. [PPV74], D Ar, Popović et al. [PPV74], ⌓ Ar, Mitin et al. [MKG75], ⚠ Ar, Ivanov et al. [IMPFD76], ▽ Xe, Ivanov et al. [IMFD76], ◇ Ne, Ivanov et al. [IMFD76], ○ H, Günther et al. [GPPR76], ⦶ Ar, Günther et al. [GLR83, GLRP77], ● Ar, Goldbach et al. [GNPP78]

the thermodynamic condition of the plasma which overlaps the region of the dc discharges for the lower level and exceeds the maximum dc power for more than one order of magnitude [BDM73, BM71, BR70b, BS72, GLR83, GLRP77, GNPP78, GPPR76, Gü70a, IMFD76, Ko71a, Ko71b, MKG75, PPV74, RG73]. On the $p - T$ plot given in figure 3.4 the pulsed-arc parameters exceed those of the stationary arcs only slightly, but they differ from them by a strong contrast in the degree of ionization and consequently also in the power dissipation. The plasmas are in LTE and may be maintained stable and stationary for periods of time long enough to record all diagnostic signals related to pressure, voltage, current, radiation etc. with high accuracy. They are homogeneous in space and form very thin cold boundary layers between the hot plasma and the walls which protect the plasma against cooling by heat conduction and the walls against melting and evaporation.

3.2.5.1. Limits of the Arc Pressure

Nonideal plasmas should have low temperatures and high charge carrier densities. The non-ideality of pulsed discharges is limited by the problem of producing high-pressure plasmas at low temperatures. One of the most important advantages of pulsed discharges over most other nonideal plasmas is their homogeneous condition in rather large volumes compared to the small boundary regions. This necessitates the ignition of the arc between rather distant electrodes. To avoid extraordinary voltages the arc must be ignited at comparatively low initial pressures and build up its plasma pressure by heating and ionizing the cold gas. The resulting arc pressure depends on the gas species, the initial pressure, the arc power, and rather sensitively on the ratio of the arc volume and the dead spaces. Of course, a high filling pressure produces a high arc pressure, but to decrease the necessary ignition voltage the electrode distance must be shortened. This increases automatically the fraction of the dead spaces of the total gas cavity whereby the fraction of particles flowing into the dead spaces increases too, and the resulting arc pressure is less than expected. Thus, the maximum pressure attainable in pulsed wall-stabilized arcs depends on the mechanical strength of the tube walls against shock waves and on the necessary ignition voltage of the gas; it ranges from 1.5 ... 2 MPa for hydrogen to 3 ... 4 MPa for rare gases.

3.2.5.2. Discharge Circuits

There are many possibilities to feed the electric power into the plasma. Before the discharge the pulse energy may be stored in magnetic fields (see, e.g., [BMM71, VKKK74, Wi70]). The introduction of this energy into the discharge volume can be established by an inductive coupling of the discharge [ABDS69a, ABDS69b, AG70c, Andr67, AVES67, MPZ70, MP69, MZP73, PMK69] or by a microwave system [Kap69]. However, the most preferred system consists of a capacitive energy storage system and a galvanic-coupled discharge volume. The advantages of this system are the good reproducibility, the simplicity compared to other methods, and the possibility to modify the duration and the shape of the pulse on a large scale. Besides, the power input of the plasma can be derived exactly from electrical measurements. This is very important for the determination of the temperature and the transport properties of the plasma.

To ensure an exact diagnostics and an uncomplicated interpretation of such experiments it proved to be very suitable to use a so-called pulse forming network (pfn) for the energy storage system whose impedance is matched to that of the discharge. This circuit (see figure 3.5) can be considered an ideal source with the voltage

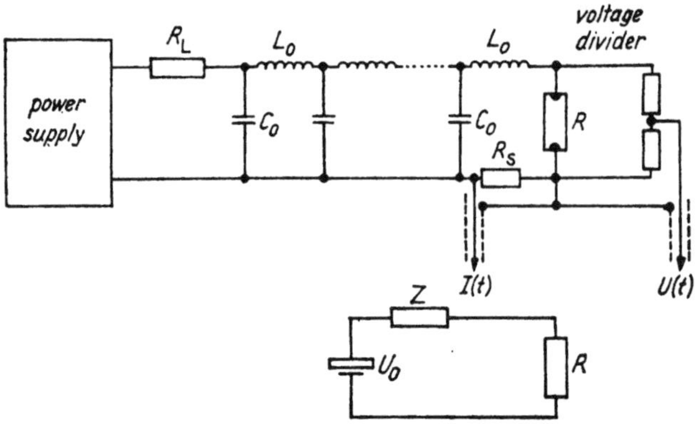

Fig. 3.5. Discharge circuit for the generation of rectangular quasi-stationary high-power pulses;
R_L load resistor, L_0 inductance of one member, C_0 capacity of one member, R discharge with the impedance R, R_S shunt

U_0 and an ohmic series resistor Z which is equal to the wave impedance

$$Z = \sqrt{\frac{L_0}{C_0}} \tag{3.20}$$

where L_0 and C_0 are the inductance and the capacitance of each of the n members of the pulse forming network. After the ignition a conducting channel forms between the electrodes whose impedance R depends on temperature and pressure the quantities of which depend on the power dissipated in the plasma. The result is a discharge with a voltage U and current I

$$U = U_0 \frac{R}{R+Z}, \qquad I = \frac{U_0}{R+Z} \tag{3.21}$$

which is stationary during a time

$$T = 2n\sqrt{L_0 C_0} \ . \tag{3.22}$$

The rise and decay time is correlated to the frequency of the circuits $\omega = (L_0 C_0)^{-1/2}$. For high-precision measurements it is useful is to tune every member of the network to fulfil the condition (3.20). This can be done by dividing every inductance into two coils whose magnetic coupling can be changed by adjusting their distance. The remaining ripple on top of the current wave form caused by the finite number of members in the pfn is much smaller than the ripple in the case of an ohmic resistor because the oscillations of the discharge current are smoothed by the feedback of the discharge current to the value of the plasma impedance [GB66]. In low-impedance discharges

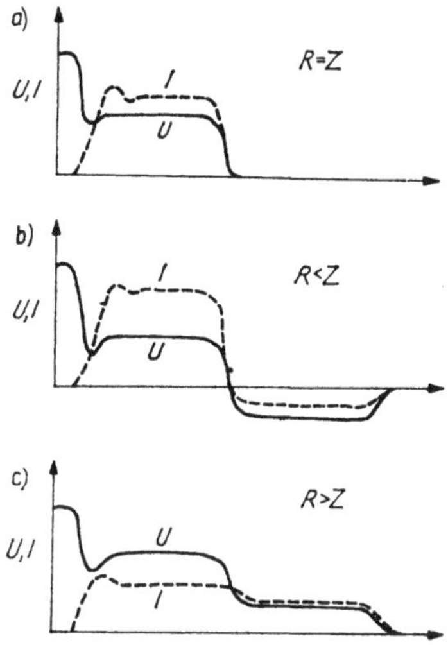

Fig.3.6. Current and voltage wave form for a matched (a)) and mismatched (b) and c)) discharge

care must be taken that the n-fold of the resistance of the coils is small compared to the impedance of the discharge R to avoid a decrease of the current during the stationary period of the pulse.

For the special case $R = Z$ the equations (3.21) read as

$$U = U_0/2\,, \qquad I = U_0/2Z\,.$$

For all other cases the circuit is mismatched and the discharge waveform described by (3.21, 3.22) is followed by one or more reflected pulses of the same length but smaller amplitudes (for schematic drawings see figure 3.6, oscillogrammes may be seen in [Gü68, Gü70b]).

3.2.5.3. Ignition and Build-Up of the Arc Column

The ignition of the discharge can be established by different methods (see figure 3.7):

(i) Pre-ionization of the gas by an rf voltage applied by a conducting electrode outside the discharge walls.

(ii) Breakdown of the discharge distance by the high voltage of a small capacitor C which is applied to the discharge device through an ignited spark gap. L_0 acts as a barrier against the short circuit of the ignition pulse across C_0.

(iii) Application of the ignition pulse by means of a transformer. The voltage may be increased by a proper transmission ratio. The capacitor C protects the pfn components against the ignition pulse.

The so-called simmer mode which keeps the discharge tube in the conducting state by means of a weak dc discharge (constricted low pressure discharge) needs an additional high power switch. It will not be described in detail here because it is used mainly for operating laser flash tubes (see, e.g., [Bec74, Lu73, SH72a]). Technical details about ignition circuits may be found in the literature [Frü65, Hoo72, Mar78, PT65]. Sometimes the delay between the ignition pulse and the beginning of the high-current discharge is of interest. It depends on the method of ignition, on the electric

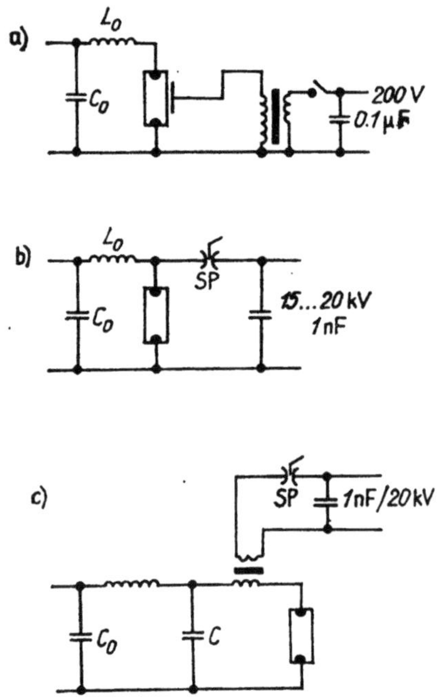

Fig. 3.7. Ignition circuits for pulsed discharges. Quantitative data for components are typical values. SP: triggered spark gap

field strength between the electrodes created by the capacitor bank, and on the filling pressure [ABSS77]. Typical values are 30 µs for a capacitive rf ignition (figure 3.7 a)) and 5 µs for a direct ignition across the electrodes by a high voltage pulse (figure 3.7 b), c)). The ignition of the discharge is followed by the build-up of the discharge channel. The character and the duration of this process depend on the ignition mode, the gas pressure and the discharge circuit, especially on the rise time of the pfn. The extension mechanisms of the thin ignition channel are explained as a combination of shock waves, thermal expansion, heating of the boundaries of the hot channel by charge carrier diffusion and by absorption of ionizing radiation [AKF73, Gü68, KF76a, Mar78, NP72, VGP71, VMNS70]. For the investigation of nonideal effects in such discharges the most interesting aspect is the time the channel takes to extend homogeneously across the discharge vessel because only after this time reproducible and consistent measurements of the plasma state and the electrical properties can be made which can be related to each other. Figure 3.8 shows as an example the relation between

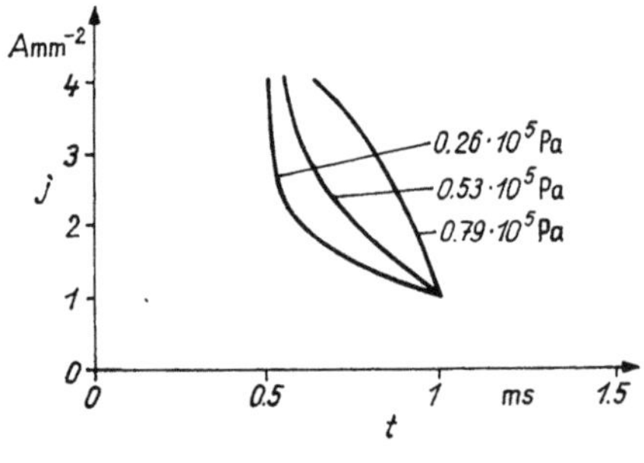

Fig. 3.8. Interdependence between the current density j during the quasi-stationary pulse and the time between ignition and the build-up of a homogeneous discharge channel for an inner diameter of 7 mm. Parameter: initial pressure [Gü68]

this build-up time and the current density during the quasi-stationary period for xenon [Gü68]. It can be seen that the minimum pulse length and the delay between ignition and measurement should be chosen carefully.

3.2.5.4. Power Ratings and Temperature

The temperature of the plasma in the stationary period is a result of the balance between heating and dissipation. As explained in section 3.1., because of the negligible contribution of heat conduction this balance can be established in a very simple form when the radiation power of the plasma as a function of temperature and pressure is known. There are numerous attempts to do this assuming some simplifications, e.g. a blackbody characteristics [Gus73, KKM74a, KR74], a grey radiation [OPZ69], or a crude introduction of the line radiation as a correction to the continuous emission [AG70c, BNRS71, BR70b, BRS74, Gü73, KKM74a, KKM75, KR74, Lit73, Rov75, Rov77]. The last calculations are in rather good agreement concerning the total values of the radiation power though they differ considerably with respect to the relation of line and continuum radiation and to the influence of radiation diffusion (see [GLR83]).

Thus, the plasma temperature is determined by the power dissipated in the plasma, and the regions of plasma conditions to be realized are determined by the maximum shock wave and thermal load which can be supplied to a discharge tube and thermal load of the wall and window materials and by the minimum electric power for establishing a stable arc. The maximum load which can be supplied to a discharge tube depends on its physical size, the pulse duration, the rise time of the current waveform, the initial pressure, and the filling gas. Data sheets of commercial flashlamp producers (see, e.g. [Pe66]) give a rough impression of the maximum ratings which lie at about 0.6 TW/m^3 arc volume for xenon plasmas at 1 ... 2 MPa. This corresponds to temperatures of 14500 and 12500 K, respectively. For hydrogen these power limits are lower; however, the corresponding temperatures are much higher: a hydrogen arc of 1 MPa produces an axis temperature of 21500 K at a power of 0.3 TW/m^3 [RG76]. Mitin et al. [MKG75] increased the power ratings using a cascade arc device in the pulsed mode. The discharge channel consisting of insulated metallic discs resists pressures of 100 MPa at specific powers of 20 TW/m^3 and allows non-idealities of $\gamma = 0.3$.

For nonideal effects low arc temperatures are of special interest. However, there is a minimum charge voltage of the capacitor bank which is necessary to develop the ignition spark to a tube filling discharge. According to figure 3.8 there is also a minimum current density required to establish a homogeneous plasma. Both these aspects correspond to a minimum power and a respective minimum plasmas temperature of the pulsed discharge which lies usually above the upper limit of steady-state arcs. This gap can be closed using a typical property of pulse forming networks: by a suitable mismatching of the wave impedance Z the plasma resistance produces reflected pulses of much less power which can be changed within large limits. Owing to the preceding strong pulse there are no problems to produce a homogeneous plasma column having a rather low temperature. Using this method the whole scale of plasma conditions from the stationary arc to high-power pulsed discharges can be established (an example is given in [GLR83]).

3.2.5.5. Discharge Tube

The apparatus for producing the test plasma should fulfil the following requirements: It should confine the plasma within a simple cylinder symmetric geometry allowing a wide range of arc lengths and diameters which can be adapted to the requirements of the spectroscopic, interferometric, and electric measurements. The ratio of arc volume and dead spaces should be as large as possible providing for high arc pressures at moderate filling pressures. The dynamic arc pressure should be measurable directly by pressure probes. In some cases wide-aperture optical windows are necessary to observe the spectral emission and its radial distribution in end-on direction. Finally, a possibility to determine the electric potential between the electrodes as well as the arc current should be provided.

It is not easy to fulfil these requirements in the same experimental arrangement. In recent years many experiments were made using simple flash lamps for laser pumping or sealed discharge tubes of similar construction. Of course, such experiments suffer from some questionable assumptions because many pieces of information as arc pressure, electric field strength, and absolute radiation intensities cannot be derived from direct measurements. Thus, special discharge devices have been developed which fulfil the requirements of the experimental techniques. Many details were derived directly or with some modifications from approved constructions [KM67, Wu58] published in the literature.

First measurements were made in cylindrical quartz tubes which were situated between two tungsten electrodes. Channels through the cathode lead to a piezoelectric pressure probe for registration of the dynamic pressure during the discharge. The radial temperature distribution which must be known for the calculation of the transport properties of the plasma was derived from side-on observations of the spectral emission. Though the distortion of the observation beam due to the cylindrical tube walls could be eliminated mathematically [Gü68] no reliable temperature distributions could be determined because of the large effect of small imperfections in the wall material. Andreev and Gavrilov [AG69b] tried to avoid this problem using discharge tubes having a rectangular cross-section, but the side-on distributions of such discharges cannot be applied to a cylindrical discharge as was shown in [Gü70b]. The only solution to this problem is an end-on observation using either a radial slit through the electrode [Gü70b] or a cylindrical quartz rod touching the plasma column in axial direction (see, e.g., [GR73, RG76]). Such discharge tubes are either L-shaped [GR73, RG76] to locate the anode outside the discharge column or, according to figure 3.9, contain a coaxial tubular tungsten anode around the cylindrical observation window **Q**. The dimensions of the window and the quartz tube are chosen for a proper optical scanning of the whole discharge cross-section. Viton seals allow a simple and fast exchange of the window when it is contaminated. The cathode contains channels to connect the piezoelectric pressure probe P with the discharge volume. Inside the tube an additional tungsten electrode with a steel or nickel ring is located which can be connected to one of the two tungsten rods R by means of a magnet. The length of this auxiliary electrode l is chosen so that

$$E \cdot l > U_a + U_c$$

where E is the axial field strength of the plasma, and U_a and U_c are the electrode

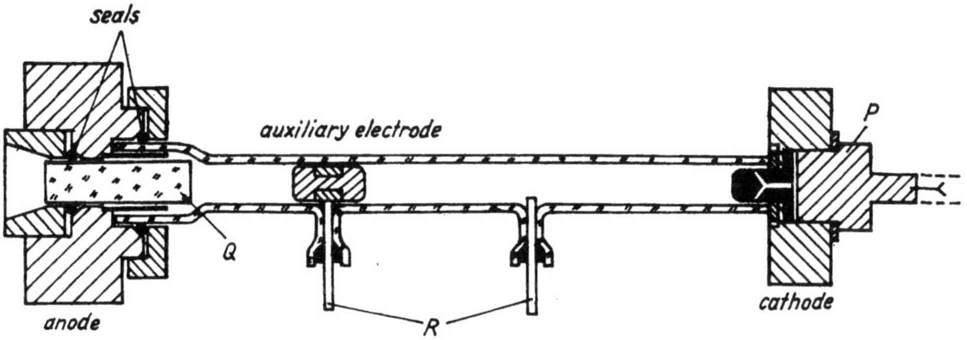

Fig. 3.9. Experimental discharge tube for the production and diagnostics of pulsed wall-stabilized cylindrical discharges;
Q quartz rod with optical quality surfaces which can be exchanged easily when contaminated, **R** electrical connections for determination of the potential of the auxiliary electrode, **P** piezoelectric pressure probe for measuring the dynamic pressure of the discharge

drop voltages. Under these conditions the arc interrupts at the position of this electrode and the total current flows through the electrode. The tungsten rod R is in contact with the electrode and gets the potential corresponding exactly to the voltage of the discharge column at this point, and we may calculate the field strength E from two oscilloscopic voltage measurements with different positions of the electrode. This technique completely avoids the problems of correction for drop voltages and the usual difficulties found when interpreting probe measurements. Obviously this procedure is justified only when the discharge parameters as plasma pressure, temperature, voltage, and current are completely independent of the position of the auxiliary electrode. It can be expected and was shown experimentally that moving the electrode does not influence the total electrode distance and the ratio of dead spaces and plasma volume and, consequently, the discharge remains completely unaffected.

A second purpose of the auxiliary electrode consists in the possibility to confine and to change arbitrarily the length of the plasma layer which contributes to the end on radiation. The use of this method for the determination of the spectral emission and absorption coefficient and of the plasma temperature is discussed in chapter 4. Figure 3.10 shows another construction which differs from the former in the dimensions of the quartz rod only. The disadvantage of this window consists in that it does not allow a radial scanning of the discharge column. On the other hand, the perturbation of the discharge plasma in consequence of this optical probe is minimum. It should be preferred for precision measurements of the axis temperature or for interferometric determination of particle number densities. In this connection it is very favourable that the active length of the plasma between the window surfaces can be chosen independently of the geometry of the discharge tube by means of a simple displacement of the quartz rods.

Generally, the choice between the two constructions on figures 3.9 and 3.10 depends on the optical path length of the plasma under investigation: In plasmas with high optical thickness the observed radiation is emitted from regions not far distant from the window, and the strong perturbation of the plasma column by the large diameter rod in figure 3.9 can result in wrong measurements. This construction supposes optical path lengths of some centimeters. Future improvements for such discharge

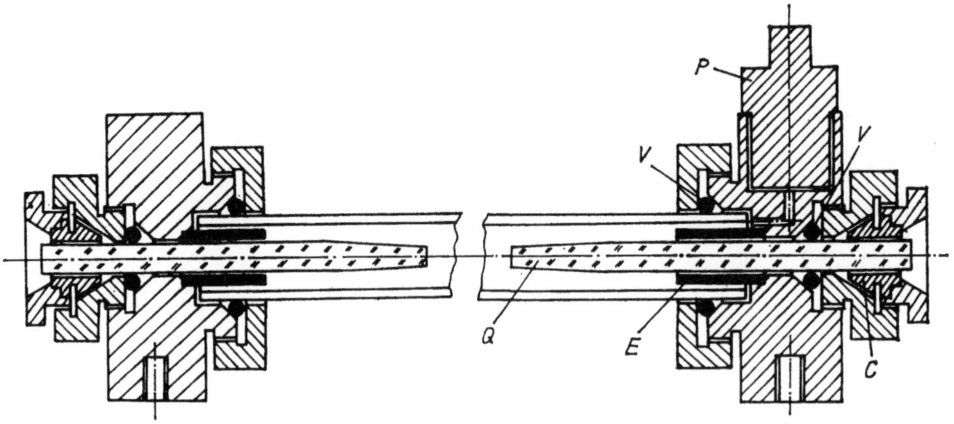

Fig. 3.10. Discharge tube for the precise spectroscopic measurement of the axis temperature and/or the interferometric determination of the particle concentration in the axis; *E* cylindrical tungsten electrode, *Q* quartz rod for end-on spectroscopy and interferometry, *V* vition seal, *P* connection for the pressure probe, *C* teflon chuck for the quartz rod

arrangements should be expected concerning the increase of the arc pressure limit in connection with improved ignition methods. Furthermore, the capability to produce high-pressure alkali plasmas is desirable in order to reach much higher degrees of non-ideality under the same precise conditions of generation and diagnostics as in rare gas or hydrogen plasmas. This development requires the availability of alkali-resistent wall and window materials as well as heat-resistent sealing technology.

4. Determination of Plasma Parameters

The information to be derived from experiments on nonideal plasmas may be divided into two different kinds: the first species is necessary to describe the thermodynamic state of the plasma under investigation, and the second one relies on the nonideal properties of this plasma such as radiation, absorption, refractive index, electrical conductivity etc. It was shown in section 1.1. that nonideal plasmas are normally in thermodynamic equilibrium with respect to the relation between neutrals and ions and to the population of energy levels. However, the strong charge carrier interaction which provides this equilibrium at the same time influences the relation between these levels and, in consequence, their population; a description of the plasma state by means of a number density of a special kind of particles as usual in the well-established arc physics becomes therefore problematic in nonideal plasmas. Furthermore, the plasma properties mentioned above which are treated in ideal plasmas with good approximation as an additive superposition of isolated elementary processes must be increasingly considered to be the results of collective effects. Thus, it is clear that we must take a careful choice in selecting both the basic plasma parameters which are suitable for an exact description of the state of the plasma and the properties of the plasma from which these parameters can be derived.

The basic properties which completely define the state of a plasma in thermodynamic equilibrium and its properties are temperature and pressure. Other parameters as electron or neutral number densities are very useful to characterize this state, but they are by no means sufficient as a basis to compare the results of theoretical considerations about nonideal effects with the experiment because these number densities themselves are a subject of nonideal plasma research and cannot be defined consistently in the case of a plasma with no negligible charge carrier interaction. The so-called quasifree electrons which are an unimportant fraction of the total number of free electrons under ideal conditions represent a considerable portion of them in a nonideal plasma. They act as free electrons or contribute to the highly excited bound states depending on the mechanism under consideration. Therefore, it is necessary to consider a particle number density derived from an observed effect of the plasma in connection with this very effect only; the number density of the same particle species will possibly change when connected with another physical mechanism. Thus, it is clear that the basic parameters temperature and pressure should not be calculated without a critical consideration of the possible uncertainties from particle number densities derived from an experiment.

4.1. Temperature

According to the preceding considerations the temperature of the plasma should be derived preferably from experimental methods which do not require the quantitative knowledge of elementary processes or particle number densities. Furthermore, it is usually not easy to apply the established methods of line spectroscopy to the diagnostics of nonideal plasmas. The intense background radiation renders the seperation of the line intensity more difficult, and it is often impossible to correct the line for optically thin radiation when the line intensity or its halfwidth is needed to determine the temperature or particle number densities. Except that, at the present time the line broadening in a nonideal plasma is an object of investigation itself rather than a diagnostic tool and new theories [Gri74] have not been compared with experimental results for high charge-carrier densities as yet. Similar problems arise in deriving plasma temperatures from line intensities or intensity ratios. Besides the perturbation of the measurements through intense background and neighbouring lines there are indications of an influence of the charge carrier density on the line intensity [Günd70, RV76] which should also be investigated experimentally in the high-density range. Considering these problems we should make use of the thermodynamic properties of the radiation in nonideal plasmas rather than the insecure knowledge of the relevant elementary processes. The calculation of the plasma temperature from an absolute measurement of the radiation intensity proved to be a very reliable procedure which should be preferred for weakly nonideal plasmas because of their comparatively high absorption. The method will be discussed in the following with special respect to the experimental conditions in weakly nonideal plasmas. The observation beam in figure 4.1 penetrates a homogeneous plasma oft he temperature T and the length l and a boundary region d. Neglecting the effect of this inhomogeneous layer, we get at first for the intensity $I_\lambda(T)$ of the emitted radiation

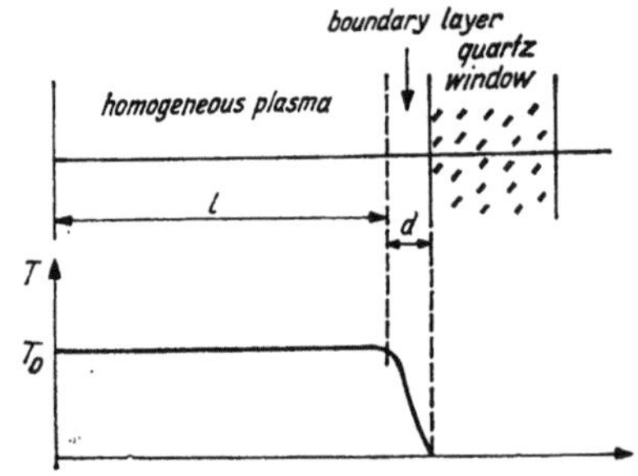

Fig. 4.1. Schematic temperature distribution along the observation beam

$$I_\lambda(T) = [1 - \exp(-\varkappa_\lambda l)]\, B_\lambda(T) \tag{4.1}$$

where $\varkappa_\lambda$ is the absorption coefficient and $B_\lambda(T)$ the blackbody intensity at the temperature T. For two measurements with different geometrical lengths l_1 and l_2 we get $\varkappa$ from

$$\frac{I_{\lambda,2}(T)}{I_{\lambda,1}(T)} = \frac{1 - \exp(-\varkappa l_2)}{1 - \exp(-\varkappa l_1)} \tag{4.2}$$

and T from the blackbody function

$$B_\lambda(T) = \frac{I_{\lambda,1}(T)}{1 - \exp(-\varkappa l_1)}\,. \tag{4.3}$$

The procedure is recommended for spatially extended plasma arrangements of constant temperature which allow a variation of the geometrical path length l within wide limits without changing the plasma state. Side-on observations in wall-stabilized or electrode-stabilized arcs as proposed by CHARNAYA [Char68] and KLEIN [Kle68] do not fulfil this condition as good as the discharge device in figure 3.9. Here the optical path length can be changed by moving the auxiliary electrode inside the quartz tube at completely constant conditions [Gü70b]. A direct change of the electrode distance in order to vary the optical path length proved to be a failure because the behaviour of the discharge changes completely: it influences the impedance of the arc as well as the ratio of discharge space to dead spaces and consequently the plasma pressure. Two consecutive measurements with different optical path lengths which are changed using this technique are by no means comparable even when the discharge current is kept constant by means of an external control.

The procedure requires at least one absolute intensity measurement. The problem consists in the large attenuation necessary to compare the plasma intensity with a radiation standard. Estimates yielded factors as large as 10^3 for the use of a carbon arc standard [MS66] which should be precise within $\approx 1\%$. It was found that one or two reflexions of the beam at a pure glas surface yielded attenuation factors between 20 and 600. For one reflexion we get from FRESNEL's law

$$r = \frac{(n-1)^2}{(n+1)^2} \cdot \left(1 + \frac{n^2 - 1}{2 - \alpha^2} \cdot \frac{\alpha^2}{n^2}\right) \tag{4.4}$$

where n is the refractive index of the glass plate. The last bracket in (4.4) is a correction for reflexion angles $\alpha \neq 0$ but $\alpha \ll 1$. It is easy to determine the refractive index n of a glass plate with an accuracy of the order of 10^{-4} after the method of ABBE from the critical angle of total reflection (see, e.g., [BW65]). The uncertainty of the reflexion factor δr depends on the error of the refractive index δn

$$\delta r = \frac{4n}{n^2 - 1} \delta n \approx 2.5 \cdot 10^{-3} \tag{4.5}$$

for $\delta n = 5 \cdot 10^{-4}$ as realized in commercial refractometers. This is much more exact than can be checked by means of photometric methods. In this way it is possible to transfer the accuracy of the carbon arc intensity calibration to the experiment. The statistical error of the method depends on the accuracy of the amplitude measurements and decreases with increasing optical thickness of the observed radiation. Assuming that $\varkappa l_1 > 2 \ldots 3$ we only get a small increase of $I_\lambda(T)$ by the correction to infinite optical path length according to (4.3), and the relative error of the temperature measurement δT is

$$\delta T = \frac{\lambda T}{c_2} \left\{1 - \exp\left(-\frac{c_2}{\lambda T}\right)\right\} \delta I_\lambda(T) \tag{4.6}$$

where c_2 is the constant of PLANCK's law. The slope of the error function $K(\lambda, T)$ shown in figure 4.2 suggests the choice of small products of wavelength λ and plasma temperature T, i.e. for a given temperature T the observation wavelength should be as short as possible. Of course, the selected wavelength depends also on other conditions, e.g. the error limits of the calibration standard as a function of the wavelength, perturbing radiation or absorption of the standard, spectral ranges with convenient absorption conditions of the plasma to be observed etc. At $\lambda = 400$ nm the error

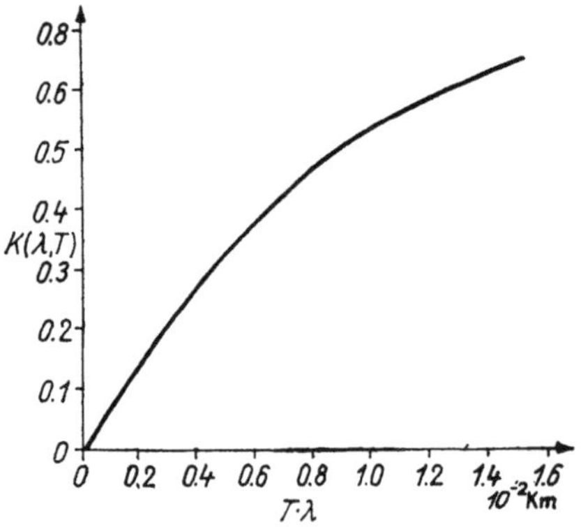

Fig. 4.2. Error function for the temperature determination from optically thick spectral intensities

limits of the carbon arc are 2% [MS66], and a total error of the absolute intensity measurement $\delta I_\lambda(T) = 5\%$ can be realized easily. The resulting temperature error of about 1% is smaller than that of usual spectroscopic temperature measurements.

Finally, the effect of the boundary layers which has been neglected as yet should be discussed. There is a transition layer between the homogeneous isothermal plasma and the cold surface of the observation window which is able to reabsorb the emitted radiation. From photographic measurements and radial scannings of the discharge channel (see, e.g., figures 3.9 and 3.10) it is known that in pulsed discharges this layer has a thickness of some tenth of a millimeter only. We will estimate the absorptivity A of this layer assuming that the dominant absorption process is due to the recombination continuum. For a linear approximation of the temperature profile given in figure 4.1 we obtain:

$$A \approx \int_0^d \varkappa(x)\,\mathrm{d}x = \frac{\mathrm{d}}{T_0}\int_0^{T_0} \varkappa(T)\,\mathrm{d}T\,. \tag{4.7}$$

The absorption coefficient $\varkappa(T)$ of the recombination continuum can be written as

$$\varkappa(T) = \varkappa_{\mathrm{at\ KU}}(\lambda, T) \cdot \xi(\lambda) \cdot n_{\mathrm{a}}(T)\,. \tag{4.8}$$

Here $\varkappa_{\mathrm{at\ KU}}(\lambda, T)$, $\xi(\lambda)$, and $n_{\mathrm{a}}(T)$ are the absorption cross-section of the photo-recombination process, the correction function according to BIBERMAN (see, e.g., [BN67, Gri74]), and the neutral particle number density, respectively. For

$$n_{\mathrm{a}} \approx \frac{p}{kT} \text{ and } \exp\left(\frac{c_2}{\lambda T}\right) \gg 1$$

we obtain

$$\varkappa(T) \approx C \cdot \exp\left(-B \cdot T_0/T\right) \tag{4.9}$$

with

$$C = \frac{e^6\lambda^3\xi(\lambda) \cdot p \cdot \gamma^*}{6\pi\sqrt{3}\, n^4\varepsilon_0^3 c^4 Z_0} = 4.78 \cdot 10^{17}\ \mathrm{Pa^{-1}\,m^{-4}} \cdot \frac{\lambda^3\xi(\lambda)\, p\gamma^*}{Z_0}$$

and

$$B = \left(\chi_{\mathrm{i}} - \frac{hc}{\lambda}\right) \cdot \frac{1}{kT_0}\,.$$

Using these simplifications (4.7) can be solved analytically:

$$A = l \cdot C\ \{\exp\left(-B\right) + B \cdot Ei\left(-B\right)\} \tag{4.10}$$

For large arguments B an expansion is recommended producing

$$A = l \cdot C \cdot \exp(-B)\left\{\frac{1}{B} - \frac{2}{B^2} + R\right\} \tag{4.11}$$

with a remainder $R < 2/B^2$.

We estimate (4.11) for a xenon plasma at $\lambda = 400$ nm, $p = 1$ MPa and $T_0 = 12000$K. With ξ (400 nm) ≈ 2 (see, e.g. [Schl68]), $\gamma^* = 6$, $Z_0 = 1$, and $l = 1$ mm we find

$$A \approx 5 \cdot 10^{-3}.$$

This result shows that all simplifications made in the preceding estimate are justified. Consequently, the uncertainty produced by reabsorption is negligible compared to the experimental error. However, for higher temperatures and/or larger pressures the absorption A increases rapidly, and this estimate should be repeated in each case.

The temperature determination method described here can be applied to the following conditions:

(i) The optical thickness of the radiation should be $\varkappa l \approx 1$ for 200 mm $\geqq l \geqq 10$ mm.

(ii) The optical thickness of the boundary layer should be negligible for the same wavelength: $\varkappa d \ll 1$.

(iii) The norm temperature of the radiation process observed should be larger than the plasma temperature and its lower level should not be the ground level; in the contrary case condition (ii) cannot be fulfilled.

(iv) The product λT should be small to diminish the experimental error.

A suitable construction of the discharge device and choice of the observation wavelength may accomplish these conditions for most weakly nonideal plasmas. The constructions described in section 3.2.4. allow a simple adaption of the geometrical dimensions to the requirements. The observation windows are extended into the plasma to suppress the absorbing gas layers and can easily be replaced when contaminated. The pulsed mode of operation prevents damage of the window surface due to overheating (see, e.g., [GR75, Gü70b]).

The method has been used successfully with the expected accuracy for rare gas plasmas as xenon [GR72, Gü70b] and argon [GLLR83, Ra74, RG75] using the recombination continua in the blue spectral range. Under some circumstances optically thick continuum radiation cannot be established even in nonideal plasmas: in the plasmas of very light gases as hydrogen or in the lighter rare gases at low temperatures. In such plasmas the optically thick radiation of strong saturated lines was used successfully. Figure 4.3 shows a spectrum of H_α from a homogeneous plasma layer of some centimeters thickness [RG76] having complete blackbody intensity at $\lambda = (625 \pm 5)$nm. The comparatively large wavelength and the high temperature produce a temperature error $\delta T \approx 2.5\%$. In argon there are strong lines in the infrared which are optically thick even at $T = 12000$ K and $p = 0.5$ MPa. Figure 4.4 shows the spectrum of an argon plasma for two geometrical observation lengths in the range 790 ... 830 nm indicating blackbody intensity in the centre of several lines [GLR83]. The temperature calculated from the intensities of these lines agrees quantitatively with that of continuum measurements in the visible. Using both radiation mechanisms a rather wide temperature range can be measured.

The large range of this temperature diagnostics is important particularly when investigating transport properties. The determination of the electrical conductivity

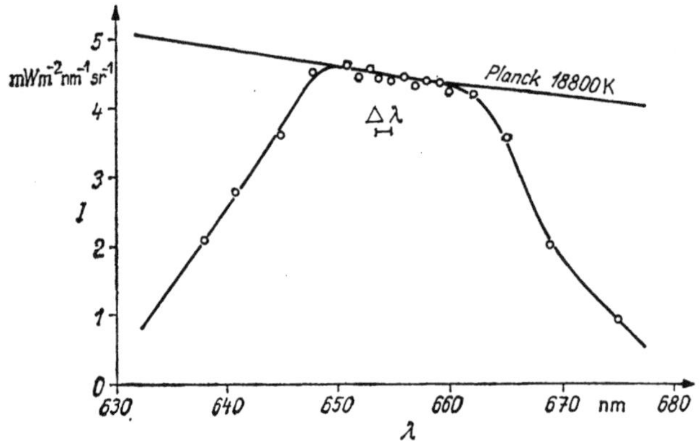

Fig. 4.3. Spectral distribution of the end-on radiation from a hydrogen plasma at $T = 18800$ K and $p = 1$ MPa

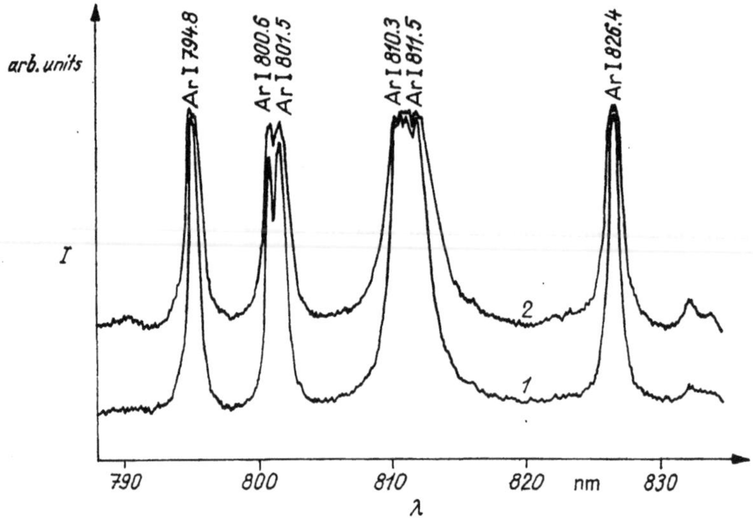

Fig. 4.4. Spectrum of an argon plasma at $T = 13450$ K and $p = 0.83$ MPa for two different optical path lengths $l = 25$ mm and $l = 70$ mm

requires the knowledge of the temperature profile of the discharge over the whole cross-section until the boundary region. The spatial resolution as well as the minimum distance of the last experimental point from the tube wall is only limited by the aperture and the length of the observation beam in the plasma when a construction of the discharge vessel according to figure 3.9 is used for end-on observation. The spatial resolution to produce a desired accuracy of the conductivity depends mainly on the temperature profile of the discharge channel. The sharp-edged profiles of pulsed high-pressure arcs proved to be not very sensitive: the temperature profiles shown in figure 3.1 (curves 2—4) yielded electrical conductivities with error limits of about 5% [GLR83, GPPR76, RG75].

In principle, temperature and conductivity profiles can also be derived from side-on observations and subsequent Abel conversions; however, the necessity to constrict the discharge by means of a material wall with high refractive index involves bad distortions of the spatial intensity distributions and a corresponding low resolution. The observation of the boundary layer seems to be nearly impossible although the beam deflection due to the cylindrical wall can be corrected in principle [Gü68]. To avoid this problem, other authors tried to get information about the radial temperature distribution from side-on measurements in arcs burning in quartz tubes having rectangular cross-sections [AG69a, AG70b]. However, the comparison of these profiles with direct end-on measurements indicates the large errors of this technique [GR72].

A complication should be mentioned that may introduce serious errors in space resolved spectroscopic measurements. As shown by Popović and Konjević [PK76], an absorption beam propagating perpendicular to a gradient of refractive index is bent in direction of increasing refraction. The radius of curvature R can be written as [BW65]

$$R = \frac{1}{n} \nabla n \cdot \boldsymbol{n}$$

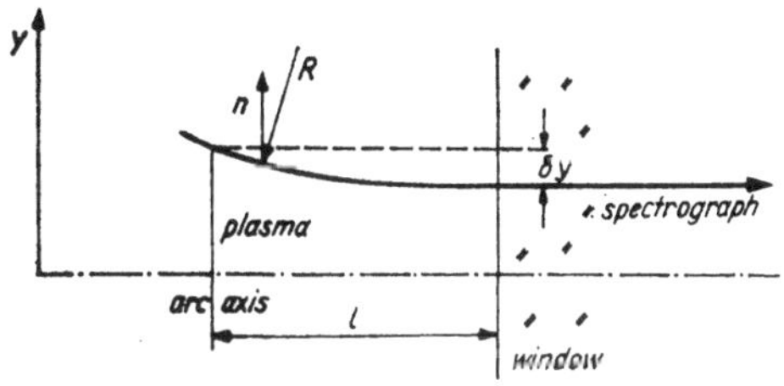

Fig. 4.5. Bending of an observation beam due to a gradient of the refractive index in a cylinder symmetric plasma

where n is the refractive index and $\boldsymbol{n}$ the unit vector in direction of the centre of curvature. Usually the path length l of the beam in the plasma is short compared to R

$$l/R \ll 1$$

and the deviation of the beam in radial direction is expressed by

$$\delta y = l^2/2R = \frac{l^2}{2n} \cdot \frac{\partial n}{\partial T} \cdot \frac{\partial T}{\partial y}. \tag{4.12}$$

This deviation can seriously influence the measured temperature profiles and consequently also the derived coefficients as conductivity. It proved to be very important for cascade arcs having small diameters and consequently large gradients [PK76], in pulsed wall-stabilized discharges of about 10 mm diameter the effect is smaller than the spatial resolution of the optical arrangement [RG76].

4.2. Pressure

The gas-kinetic pressure is the second of the basic plasma parameters which are directly accessible to the experimental measurement. In stationary discharges its determination is no problem in contrast to pulsed plasmas. Therefore, many authors calculated the pressure during a pulsed discharge from Dalton's law under the assumption that the number density of heavy particles remains constant (see, e.g., [ABD70, AG70a, AG70c, AG71, CSL68, VMNS70]). However, this pressure deviates substantially from the pressure determined by direct measurements because the random gas flows into dead spaces and cold boundary layers cannot be taken into account. The physical conclusions drawn in the mentioned papers should be considered carefully for this reason. ANDREEV himself corrected his procedure in a later paper [AG71]. The usual detectors for transient pressures are piezoresistive or piezoelectric probes. They are stable and linear components with fast response from *dc* until $10^6\ s^{-1}$, so they can be calibrated stationarily and follow the pressure signal with rise times $<10^{-5}$ s when they are in good connection with the plasma [Gü68]. Figures 3.9 and 3.10 show approved constructions which provide also protection against thermal stress of the probe.

4.3. Particle Number Densities

It is well known that the conventional diagnostic methods (line shape measurements, continuum intensities) can no longer be considered independent tools for the determination of particle number densities in high-pressure plasmas. For instance, at electron number densities above $10^{24}\ m^{-3}$ the measured line shape cannot unambiguously be related to calculated profiles because the intense continuum level which must be subtracted from the line intensity is itself under discussion in nonideal plasmas and/or large corrections from neighbouring lines must be applied to the measurements as shown by PRESTON [Pr77] on the basis of a computer based model. Unfortunately, this picture is confused still more by the fact that a tendency of line intensities to decrease towards larger electron densities may be present in the experiment (GÜNDEL [Günd70], WEISHEIT and SHORE [WS74], RUZDJAK and VUJNOVIĆ [RV76], GUROVICH and ENGELSHT [GE77]).

Similar difficulties arise in deriving electron densities from the measurement of the continuous emission coefficient ε_λ

$$\varepsilon_\lambda \sim \frac{n_e^2(T)}{\lambda^2 T^{1/2}} \cdot \xi(\lambda) \,. \tag{4.13}$$

The application of (4.13) is problematic in the case of high densities because the ξ-factor itself becomes a function of the plasma parameters as may be concluded from both experimental and theoretical studies [BOS79, GNP77, KK78, KKN77, MW76, No79]. Moreover, under such conditions strongly broadened lines and the quasi-continuum in the neighbourhood of series limits may influence the recombination continuum considerably. Several attempts have been made to include high-density corrections in the calculation of the continuous emission coefficient (see, e.g., [BN67, GZKM71, KKN77, No79]). However, these calculations differ widely depending on the level of approximation and have partially resulted in contradictory conclusions. Thus, the same arguments that apply to traditional line spectroscopic methods also exclude the application of continuum measurements to the determination of particle densities in nonideal plasmas.

As a result of limits set to conventional spectroscopic techniques another class of diagnostics based on analyzing the interaction of a known external radiation with the plasma (absorption and scattering of radiation, variation of phase relations) has been employed to evaluate particle number densities. Among them the laser interferometry (see, e.g., [BK75, BRS69, DGRU77, DP73, HK67, WT76]) and the schlieren technique [GVC78] have been developed as reliable diagnostic tools even at elevated electron densities. The former method is usually based on recording the temporal change of the refraction index Δn corresponding to a simultaneous change of plasma parameters. Δn is calculated from the number of fringes K created during the rise or decay period of the plasma

$$\Delta n = \frac{K \cdot \lambda_L}{L}, \tag{4.14}$$

λ_L being the wavelength of observation and L the plasma length penetrated by the laser beam. Weiss and Teuber [WT76] refined this technique taking into account not only the number of fringes as integer but also fractions of it by measuring the phase relation of the interference signal. This phase information is derived analytically from the amplitudes of two interference signals which are phase shifted to each other by $\pi/2$. In plasmas of higher density the experimental conditions are disturbed by a considerable absorption of the laser radiation. Therefore the optical path length through the plasma should be rather short, and a high accuracy in refractive index requires an exact analysis of the phase information. Furthermore, the high and time dependent absorption causes an additional amplitude modulation of the interference signals. The method used by Weiss and Teuber would introduce an error in the resulting refraction index which can hardly be estimated. For this reason Djordjevic et al. [DGRU77] used a graphical method to determine the phase of the interference signal that automatically eliminates amplitude modulations of the laser intensity due to absorption. The optical set-up is schematically shown in figure 4.6. A fit adjustment

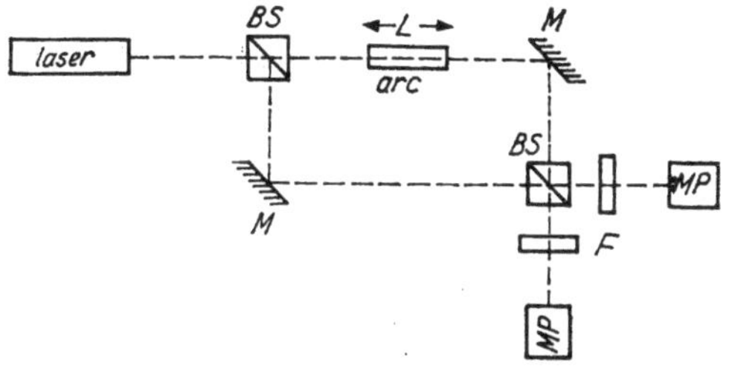

Fig. 4.6. Schematic drawing of an Mach-Zehnder interferometer with beam splitters **BS**, mirrors **M**, narrow band filters **F** and multipliers **MP**

of the interferometer mirrors produced fringe patterns behind the beam splitter from which suitable phase points could be selected and projected on both multipliers. These phase-shifted signals from the multipliers were fed to the deflection plates of an xy-oscilloscope and produced a number of more or less distorted circles on the screen the centre of which was shifted during the processes due to absorption of the laser radiation and to the emision of the plasma. Figures 4.7 a) and b) show the diagnostics of a wall-stabilized xenon pulsed discharge with a rectangular current pulse of 1 ms duration. The shift of fringes corresponds to a change of plasma parameters from the stationary state maintained during the last part of the pulse to the cold gas. Setting $K = K' + (\alpha + \beta)/2\pi$ where K' is the number of full circles and α, β are angles as defined in figure 4.7 b), the corresponding change of refraction index may

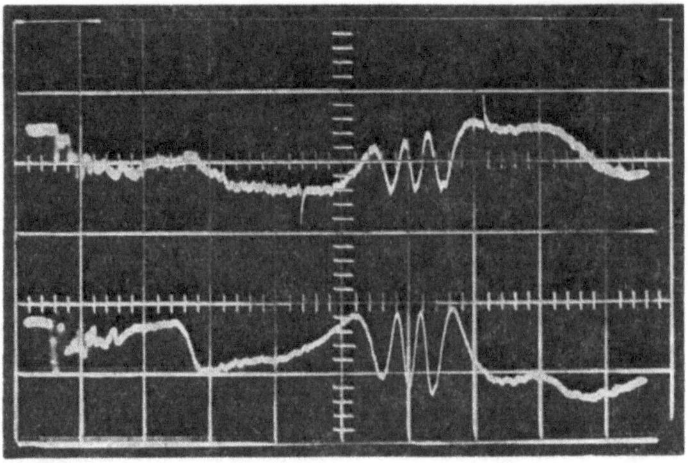

Fig. 4.7a) Interference signal at $\lambda_L = 1.15\,\mu m$; the needles on the upper signal mark the limits of registration for the polar diagram (time base: 0.2 ms/div)

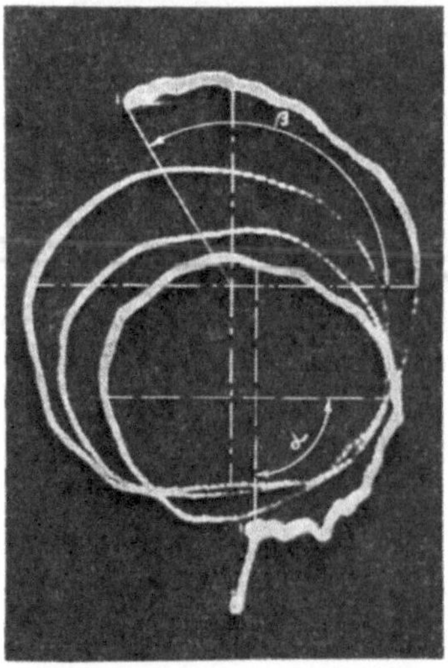

Fig. 4.7b) Polar diagram of interference signals

be calculated from (4.14). Using this technique, the accuracy in counting fringes was estimated to be about 1/50 which is comparable to the accuracy reached in transparent plasmas [BK75]. An additional uncertainty is associated with the determination of the plasma length penetrated by the laser beam. To avoid the tedious and delicate procedure of the Abel inversion, many measurements were carried out end-on neglecting inhomogeneities near the electrodes and/or effects of cold boundary layers. In this manner the path L is affected by a systematic error the magnitude of which can hardly be estimated. However, using a plasma configuration that avoids dead spaces and inhomogeneities in direction of observation — a suitable arrangement fulfilling these requirements is shown in figure 3.10 — it may be assumed not to exceed 1%.

The capability of the schlieren technique for the diagnostics of stationary high-pressure arcs has been demonstrated in [GVC78]. This method is based on recording the deflection of a light beam travelling through a plasma with a density gradient ∇n_j. It has the advantage of requiring only a simple optical set-up, but obviously this technique is qualified only for cascade arcs having small diameters and consequently large gradients over the whole cross-section. In pulsed wall-stabilized discharges with flat radial distributions this effect would chiefly be due to the gradients in the vicinity of the tube walls, and it is difficult to relate a measured beam deflection to the axis electron density.

The evaluation of particle number densities from fringe shift or schlieren measurements requires the knowledge of the plasma refractive index as a function of number densities. Often this relationship can be written as follows

$$n = \sum F_j n_j \tag{4.15}$$

where the coefficients F_j express the dispersion properties of the corresponding components. They are different both in sign and wavelength dependence.

The major contribution to n comes from the interaction of the laser radiation with the free charge carriers. As shown by HERMANSDORFER [He68b], the inclusion of $e - i$ and $e - a$ collisions in this interaction process leads to the following expression for the complex refractive index of free electrons

$$n_{\text{el}}^2 = 1 - \frac{(\omega_{\text{p}}/\omega_{\text{L}})^2}{h + ig\,\langle\nu\rangle/\omega_{\text{L}}}, \tag{4.16}$$

$\omega_{\text{p}} = (n_{\text{e}}e^2/m_{\text{e}}\varepsilon_0)^{1/2}$ being the electron plasma frequency, ω_{L} the frequency of observation, $\langle\nu\rangle$ the effective collision frequency

$$\langle\nu\rangle = -\frac{4\pi}{3n_{\text{e}}}\int_0^\infty v^0(\nu_{\text{ei}} + \nu_{\text{ea}})\,\frac{\partial f_0}{\partial v}\,\text{d}v\,, \tag{4.17}$$

and h, g correction functions according to SHKAROFSKY [SJB66] which depend on the ratio of electron-atom and electron-ion particle collision frequencies as well as on the ratio of effective collision frequency and frequency of observation.

If we restrict the electron density to the range below 10^{25} m^{-3}, the ratio $\langle\nu\rangle/\omega_{\text{L}}$ proves to be $\lesssim 10^{-3}$ for all frequencies ω_{L} usually applied in diagnostics, and because of $g = h = 1$ for $\langle\nu\rangle/\omega_{\text{L}} \ll 1$ the simpler relation for the refractive index

$$n_{\text{el}}^2 = 1 - \left(\frac{\omega_{\text{p}}}{\omega_{\text{L}}}\right)^2 \tag{4.18}$$

can be used.

Further contributions due to the interaction of the laser radiation with bound electrons can be deduced from the so-called KRAMERS-KRONIG relation (see, e.g., [LL74]) that describes the connection between the absorption coefficient $\varkappa(\omega)$ and the refractive index

$$n^2(\omega_{\text{L}}) = 1 + \frac{2c}{\pi}\,⨍_0^\infty \frac{\varkappa(\omega)\,\text{d}\omega}{\omega^2 - \omega_{\text{L}}^2}\,. \tag{4.19}$$

Taking bound-bound and bound-free transitions into account for the absorption coefficient, (4.19) represents the general expression for the refractive index of neutral particles. The evaluation of the integral requires the knowledge of the oscillator strengths and the absorption line profiles for bound-bound contributions and the absorption cross-sections in the case of bound-free transitions. In earlier works the refractive index of the cold gas was used, i.e. only the resonance transitions were considered in $\varkappa$ [DP73, BRS69]. Baum et al. [BHU75] calculated F_a for argon taking into account excited states of the neutral atom. The comparison between their calculation and experiments in cascade arcs, however, yielded considerable discrepancies. Recently [RGU80] the above expression was analyzed for a weakly nonideal hydrogen plasma including high-density corrections both for bound-bound and bound-free transitions. In figure 4.8 the results are shown for two wavelengths of

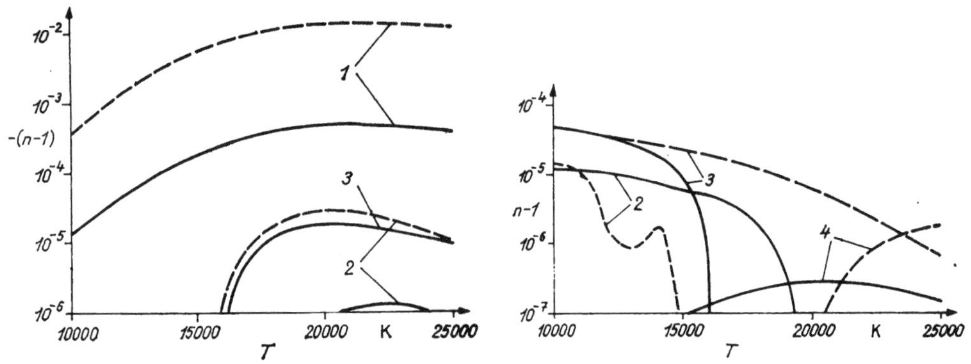

Fig. 4.8. Refractive index for a hydrogen plasma with $p = 2\,\text{MPa}$ at $\lambda_L = 0.63\,\mu\text{m}$ (———) and $3.39\,\mu\text{m}$ (— — —); due to free electrons: (1); due to atoms: (2) — boundfree transitions, (3) — bound-bond transitions, (4) — quasi-continuum

the He-Ne laser. It can be seen that for $\omega_L = 0.63\,\mu\text{m}$ the atoms contribute to the refractive index more or less in the whole temperature range, whereas at a wavelength of $3.39\,\mu\text{m}$ this contribution is always small compared to that from electrons. In particular, there is not such a great influence of excited states as indicated by Baum et al. [BHU75] if the induced emission is included and the number of levels which contribute to the refractivity is limited on account of screening by the surrounding plasma.

The coefficient F_i, finally, describes the dispersion properties of the ion. It contains a component similar to (4.18) which can be neglected because of the large ion mass and a component according to (4.19) which should not be neglected without detailed considerations.

If the electron concentration is not too high and particle interactions can be neglected in the expression for n_{el}, the refractive index is not very sensitive to non-ideality effects. In that way, IR interferometry seems currently to be the most reliable method for the determination of electron densities even at higher pressures. Taking into consideration possible uncertainties due to quasi-bound particle states, an accuracy $\delta n_e = 2 \ldots 3\%$ can be achieved with suitable plasma sources and an appropriate detection of interference signals. The number density of neutral particles can be obtained in the usual way by measuring fringe shifts at a second wavelength. However, due to the weak dependence of the plasma refractivity on this species a probable accuracy of no better than $15 \ldots 20\%$ is indicated.

It is worth noting that no radiation of frequencies smaller than ω_p can be transmitted through a medium whose refractive index is given by the expression (4.18). For $\omega < \omega_p$ namely, n_{el} becomes imaginary and the plasma exhibits properties which resemble closely the phenomenon of total reflection in normal optics (see, e.g. [Os60]). However, if collisions are taken into account according to (4.16), radiation can in fact propagate beyond this limit, but strong attenuation occurs around $\omega = \omega_p$. Skowronek et al. [SRGC70] made use of this phenomenon to determine the concentration of electrons in an exploding gaseous filament. The drop in the spectral brightness observed experimentally in the near infrared was interpreted as being due to the plasma frequency with a corresponding electron density of 10^{27} m^{-3}. However, regarding the moderate accuracy involved in this procedure — an uncertainty in the determination of ω_p implies twice as large an uncertainty for n_e — the diagnostics by means of brightness measurements seems to be difficult.

Until today, only few experimental studies have been reported which use laser light scattering for diagnostic purposes of high-density low-temperature plasmas. This may be explained by restrictions with respect to the interpretation of the scattering spectra as well as by experimental difficulties which have to be overcome before this technique can be developed into a reliable diagnostic tool. Specifically, the plasma refractivity may raise some problems in localizing the scattering centre and/or the large amount of radiation from the plasma can deteriorate the signal-noise ratio. Moreover, as can be concluded from the estimates [BPMM77, Py75], the range of parameters where laser scattering can be applied successfully cannot easily be extended by using higher power laser sources because perturbation effects may become important. For instance, if a temperature $T = 10^4$ K is assumed, the limit of electron concentration up to which THOMSON scattering is possible lies between 10^{23} and 10^{25} m^{-3} depending on the power of the probing laser and the mode of operation. Recent experimental studies on stationary cascade arcs [PBZM78] have demonstrated the capability of laser scattering for the determination of plasma parameters. However, at the present time this technique cannot be considered to be an alternative to well-established interferometric methods because of the relatively low precision even at less restrictive low-pressure conditions ($\delta T \lessapprox 25\%$, $\delta n_e \lessapprox 20\%$).

Another method to evaluate particle number densities are absorption measurements. However, it is usually not easy to apply such type of experiments to the diagnostics of nonideal plasmas since there are indications of an influence of high-density corrections on the absorption coefficient. GLASSER and CHAPELLE [GC79] actually conclude by theoretical considerations that the infrared free-free continuum absorption should exhibit some peculiarities depending on the type of small $\ln \Lambda$ corrections applied. The results in [GC79] are given in the form of a GAUNT factor

$$G = \frac{\sqrt{3}}{\pi} \cdot \frac{\langle \nu_{ei} \rangle}{\langle \nu_0 \rangle} \ln \Lambda \frac{u}{1 - e^{-u}} \tag{4.20a}$$

where $u = h\nu/kT$, $\langle \nu_{ei} \rangle$ is the effective electron-ion collision frequency (2.39), and $\langle \nu_0 \rangle$ is the value of $\langle \nu_{ei} \rangle$ in the $\ln \Lambda \gg 1$ case. Note that Λ is a function of the effective screening length which is under discussion in nonideal plasmas. Taking for a basis a COULOMB potential with an impact parameter cut-off $\langle \nu_{ei} \rangle$ is given by the expressions (2.57). However, in their calculations, GLASSER and CHAPELLE, using the $\Lambda \gg 1$

expression of $\langle \nu_{ei} \rangle$, find

$$G = \frac{\sqrt{3}}{\pi} \left(\ln \Lambda - \ln \left(\frac{3}{2} \right) - 0.577 \right) \frac{u}{1 - e^{-u}} \tag{4.20b}$$

which cannot be applied in the domain $\ln \Lambda < 3$ since further corrections become important according to (2.57). It will be mentioned in this connection that the authors of [GC79] have estimated the non-ideality effect on the free-free GAUNT factor assuming an enhancement of the cutting length as indicated by (2.82). However, the calculations have partially resulted in contradictory conclusions because instead of (4.20b) the $\ln \Lambda \gg 1$ expression of G was taken as a reference.

Sometimes X-ray absorption measurements have been applied to estimate the density n_h of heavy particles in strongly nonideal plasmas (see, e.g., [BGDF75, GKKM76, IMFD76, MKG75]). Owing to the high energy of photons, this diagnostics is applicable even in opaque dense plasmas. Concerning the uncertainties involved in the measurements, however, there are different estimates ($\delta n_h \approx 5 \dots 40\%$ according to [KMG73, LF71]).

In many cases the pressure is known from experimental conditions (e.g. stationary cascade arcs, pressurized containers) or can be measured directly by means of a probe. Using the temperature determined spectroscopically (see section 4.1.1.), particle densities may be derived from both quantities via the Saha equation and the equation of state assuming LTE conditions and the validity of quasi-neutrality. Concerning the high-density corrections that must be applied in performing such equilibrium composition calculations the reader is referred to section 2.1.

5. Evaluation of Transport Properties from the Experiment

Many experimental data on the electric properties of weakly nonideal plasmas are available from the literature of the last years. These experiments can roughly be divided into two groups concerning their intentions: On the one hand, it was necessary to investigate the electrical characteristics of pulsed discharges in a wide field of discharge parameters for many technical applications regardless of the plasma conditions or the acting elementary processes. The contradictions between the different published results and their limits should be discussed. The second group connects the electrical data to the plasma state to compare them with the results of the theory of transport phenomena. The published transport coefficients and conductivities will be discussed with respect to the diagnostic problems mentioned in chapter 4. as far as they can influence the results, especially assumptions about the plasma pressure, the particle number densities derived from interferometric measurements, and about the radial profiles of electrical discharges. Some authors calculate the cross-section for charge-carrier interactions from the electric data either neglecting the effect of collisions with neutrals in the ground and excited states or introducing them into the formalism on a very elementary level. Such simplifications can lead to numerical deviations and wrong conclusions or limit the validity of the derived expressions. Procedures avoiding these sources of error as far as possible are described. The comparison of theory with the experimental results derived following these procedures yields corrections which can be related to nonideality effects. Taking into account these corrections the electrical properties of weakly nonideal plasmas can be calculated with high accuracy within wide ranges of temperature, pressure, and ionization degree.

5.1. Derivation of the Electrical Conductivity from Experimental Data[1]

5.1.1. Integral Method for the Evaluation of the Electrical Conductivity

The integral method for the evaluation of electron transport properties from basic experimental results substituted in some of the balance equations was described by Anderson [Ande67, Ande68]. This is a consequent procedure with well defined and controlled numerical accuracy but with high demands in accuracy of basic experi-

[1] The results of this section have been obtained in cooperation with S. S. Popović.

mental information, such as voltampere characteristics, axis temperature, and radial temperature profiles. The last feature makes the procedure inferior to methods developed seperately for a particular transport coefficient, although it gives as a final result a complete set of thermophysical properties for the observed ionized gas.

The procedure starts from the ELENBAAS-HELLER equation

$$\frac{1}{r} \cdot \frac{\mathrm{d}}{\mathrm{d}r}\left(r\lambda \frac{\mathrm{d}T}{\mathrm{d}r}\right) + \sigma E^2 - S = 0 \tag{5.1}$$

and Ohm's law

$$I = 2\pi E \int_0^R \sigma(r)\, r\, \mathrm{d}r \tag{5.2}$$

for an axisymmetric plasma column. In the above expressions T is the temperature, λ is the thermal conductivity, σ is the electrical conductivity, E is the electric field, S is the radiated power per unit volume, I is the arc current and R is the arc radius.

Usually it is assumed that the axial component of the field strength is constant over the whole cross-section of the arc, except for a small layer close to the wall. This problem is, however, of no significance in arcs with walls made from dielectric materials.

In (5.2) the unknown quantity is the electrical conductivity σ which is basically a function of temperature. So, (5.2) is an integral equation but for the sake of the numerical procedure for its solution it should be modified to get a standard form. This is usually done by substituting $x = r/R$ and by presenting the measured set of inverse temperature profiles in the form $x = x(T, T_{\mathrm{ax}})$ [SP62]. These modifications are of course justified only when the temperature profiles $T(x)$ are monotonously decraesing functions and the effect of the wall temperature T_0 is negligible. The relation (5.2) can now be written in the following form

$$\frac{I}{\pi R^2 E} = \int_{T_0}^{T_{\mathrm{ax}}} \sigma(T)\, g(T, T_{\mathrm{ax}})\, \mathrm{d}t \tag{5.3}$$

where $g(T, T_{\mathrm{ax}}) = -\frac{\partial x}{\partial T}$. By differentiation with respect to T_{ax} one gets

$$\frac{1}{\pi R^2} \frac{\mathrm{d}}{\mathrm{d}T_{\mathrm{ax}}}\left(\frac{I}{E}\right) = \sigma(T_{\mathrm{ax}})\, g(T_{\mathrm{ax}}, T_{\mathrm{ax}}) + \int_{T_0}^{T_{\mathrm{ax}}} \sigma(T) \frac{\partial g(T, T_{\mathrm{ax}})}{\partial T_{\mathrm{ax}}}\, \mathrm{d}T \tag{5.4}$$

and, finally,

$$\sigma(T_{\mathrm{ax}}) + \int_{T_0}^{T_{\mathrm{ax}}} K(T, T_{\mathrm{ax}})\, \sigma(T)\, \mathrm{d}T = f(T_{\mathrm{ax}}) \tag{5.5}$$

where we identify as the inhomogeneous part

$$f(T_{\mathrm{ax}}) = \frac{1}{\pi R^2 g(T_{\mathrm{ax}}, T_{\mathrm{ax}})} \frac{\mathrm{d}}{\mathrm{d}T_{\mathrm{ax}}}\left(\frac{I}{E}\right) \tag{5.6}$$

and as the kernel

$$K(T, T_{\mathrm{ax}}) = \frac{1}{g(T_{\mathrm{ax}}, T_{\mathrm{ax}})} \frac{\partial g(T, T_{\mathrm{ax}})}{\partial T_{\mathrm{ax}}} \tag{5.7}$$

and $\sigma(T)$ as an unknown function of an inhomogeneous VOLTERRA equation of the second kind. This equation has a convergent numerical solution in the form [Ande67]:

$$\sigma(T) = \sum_{n=0}^{\infty} \varphi_n(T) , \tag{5.8a}$$

$$\varphi_0(T) = f(T) , \tag{5.8b}$$

$$\varphi_{n+1}(T) = -\int_{T_0}^{T} K(T', T)\, \varphi_n(T')\, dT' . \tag{5.8c}$$

An integral equation of the same type can be obtained for the thermal conductivity function $\lambda(T)$ by transforming (5.1) [Ande68]. The unfavourable fact is that an interval around T_{ax} should be used to construct $\lambda(T)$, so the range of application is reduced again, since for a flat temperature profile the convergence is poor.

Basic experimental information required for this procedure are the voltampere characteristics

$$\frac{I}{E} = f(T_{ax}) \tag{5.9a}$$

and the inverse temperature profiles

$$x = \frac{1}{R^2} [r(T, T_{ax})]^2 . \tag{5.9b}$$

So, it is necessary to perform a large scale of accurate temperature, electric current and field strength measurements in a wide range of supplied electric power to avoid errors during the smoothing procedure. A particular problem is the construction of the kernel function $K(T', T)$ which requires a large number of temperature profiles. The procedure cannot be applied when the temperature profiles $T(x)$ are not monotonously decreasing functions. Flat temperature profiles that can be found in high-pressure heavy noble ionized gases can produce serious convergence difficulties.

These are some of the reasons which made this method in its integral form practically inapplicable at the present stage of experimental techniques. However, in some cases, when the radiation term in (5.1) can be presented by a temperature gradient, the unknown function $\sigma(T)$ can be determined by simplified procedures. These methods were developed by PLANTIKOW [Pl69] and UHLENBUSCH [HMU72].

5.1.2. Trial Function Methods

In high-current pulsed arcs the energy balance is dominated by emission of radiation, and the transport coefficients obtained using the procedures of PLANTIKOW or UHLENBUSCH sometimes would not be unique functions of temperature. In the case of radiation-dominated arcs (5.3) has to be solved in a different way without using the energy balance.

Various methods have been developed for this purpose, most of them based on a priori assumptions on the temperature dependence of the electrical conductivity. Some arbitrary constants should then be optimized so that (5.3) is satisfied.

In some earlier papers the electrical conductivity function was represented by a piecewise broken linear dependence [MRV70, SP62]. The function $\sigma(T)$ was constructed on the basis of experimental temperature profiles at successively increasing values of current starting from the lowest axis temperature. The accuracy of this method depends on the number of linear steps, so the measurements should be performed at close intervals of arc currents. HERMANN and SCHADE [HS70] improved the piecewise linear function method by introducing a polynomial expansion of the measured radial temperature distributions. A further improvement was introduced by DEVOTO and MUKHERJEE [DM73] with a trial function

$$\sigma(T) = aT^{-b} \exp\left(-\frac{c}{T}\right). \tag{5.10}$$

The parameters a, b, and c were adjusted so that (5.2) with measured temperature profiles, arc currents, and elecrtic fields was satisfied in each set of measurements.

In all quoted methods the trial function $\sigma(T)$ was introduced formally with adjustable constants without physical meaning. In order to overcome this disadvantage, attempts were made in earlier papers to solve (5.3) for xenon and argon pulsed arcs using a rather simple temperature dependence [GR72, RG75]

$$\sigma = AT^{3/2} \tag{5.11}$$

inspired by SPITZER's formula (2.37). Using experimental temperature profiles $T(r)$ one can get a conductivity profile $\sigma(r)$ and determine the constant A to fit (5.2) with experimental voltampere characteristics. In that way, A gives an information on the COULOMB logarithm which is incorporated in the SPITZER expression. This approach is justified only when the arc plasma has a very high degree of ionization. Even then the significance of the information obtained is rather doubtful because the Spitzer expression is a very simplified one.

This simple expression of a trial function can be applied only when the influence of neutral particles can be neglected which is not the case at lower temperatures. The effect of neutral particles is that the dependence $\sigma(T)$ is much stronger than (5.11), So, the trial function should be introduced by the expression (5.11) either in a piecewise form again or by defining a stronger temperature dependence for the conductivity.

In some recent papers [GLR83, GPPR76] the electrical conductivity trial function $\sigma(T)$ was constructed in the same way as the theoretical conductivities for ideal plasmas were obtained. The details of the procedure for the evaluation of the ideal plasma electrical conductivity are given in chapter 2. and here we will only stress some relevant points.

The final result of this procedure is the expression of electrical conductivity as given by (2.21) or (2.40). Both formulas give a correct description of the influence of all kinds of particles on the electrical conductivity. It is important to underline here that the interaction of the electrons with neutrals in the ground as well as in excited states can be introduced without contradiction within the frame of binary collision approximation, as was shown in section 2.2.3.2. This is an opportunity to eliminate the influence of neutral particles in every particular experimental situation in order to study non-ideality effects on the COULOMB part of electrical conductivity.

The interaction between charged particles was described in detail in chapter 2. and it is not necessary to treat it here. The electron-ion collision cross-section which is introduced in the trial function $\sigma(T)$ has the form given by (2.54) with a free parameter x in the COULOMB logarithm

$$\Lambda = \frac{12\pi\varepsilon_0 r_s kT}{e^2} = \frac{12\pi\varepsilon_0 DkT}{e^2} x \tag{5.12}$$

where r_s is an effective screening radius and D the DEBYE length.

With the radial temperature distribution and pressure obtained by experiment one can easily construct the trial function $\sigma(T)$. The parameter x is to be changed in an iterative procedure until the relation (5.2) with experimental values for electric current and field strength is satisfied.

As was already mentioned, the plasma composition and influences of neutral particles should be treated as exactly as possible and a satisfactory procedure is described in sections 2.1. and 2.2.3.2.

The choice of the effective screening radius is the only arbitrary step in this treatment and it has a direct influence on the electrical conductivity function $\sigma(T)$. With the free parameter x as defined by (5.12) the conductivity function $\sigma(T)$ is completed.

Besides, the experimental error for obtained x values is rather large. It can be estimated by means of the simple Spitzer formula

$$\sigma = C \cdot \frac{T^{3/2}}{\ln x\Lambda} . \tag{5.13}$$

Then the relative error in x is approximately

$$\delta x \approx \ln (\Lambda x)\, \delta\sigma + \frac{3}{2} \ln (\Lambda x)\, \delta T . \tag{5.14}$$

With typical values $\ln \Lambda = 3$, $x = 2$, $\delta\sigma = 6\%$, $\delta T = 1.5\%$ one gets $\delta x \doteq 25\%$.

Even with all experimental efforts described here, the accuracy of this result is not completely satisfying, but it can hardly be improved. Nevertheless the experimental value is exact enough to estimate the behaviour of x at small $\ln \Lambda$.

So, besides the data on electrical conductivity at high densities, we can get rather reliable information on the interaction between charged particles which can serve as an experimental basis for a dense plasma theory. Performing the procedure described here using experimental data derived from plasmas of different elements and different thermodynamic composition, from the deviations or agreements of the x values for plasmas at similar non-ideality, conclusions can be drawn about the consistency of the model.

5.2. Electrical Characteristics of Pulsed Discharges

The numerous technological applications of discharges in weakly nonideal plasmas require the design and the optimization of the discharge circuits. In the last ten years the lack of reliable knowledge about the electrical properties of such plasmas has led to many empirical models and formulas describing the connection between voltage and current. The assumption of a homogeneous distribution of the plasma density and temperature across the discharge — which is not justified in all cases (see sections 3.1.2. and 3.1.3.) — results in a set of proposed formulas predicting the connection between the electrical field strength E, the electrical conductivity σ, and the current density i in the form

$$E = Ai^n \tag{5.15}$$

or

$$\sigma = Bi^m\,. \tag{5.16}$$

Table 5.1 lists a selection of coefficients A and n according to (5.15) for linear xenon pulse discharge tubes proposed by different authors. The coefficients are derived from experimental current and voltage measurements made with closed flash tubes

Table 5.1. Proposed electrical characteristics of xenon discharge tubes
Coefficients A and n refer to (5.15) where E is in V/m and i in A/mm². $\varnothing_i$ inner tube diameter, l_e electrode distance

notation in figure 5.1	author	reference	A	n	conditions
a	Marshak	[Mar78]	$4.7 \cdot 10^2$	2/3	$\varnothing_i = 0.85 \ldots 7$ mm
b	Demenik et al.	[DMSF 68]	$8 \cdot 10^2$	1/2	$\varnothing_i = 11 \ldots 36$ mm $l_e = 240 \ldots 900$ mm $i = 10 \ldots 80$ A/mm²
				5/11	dissipation: blackbody radiation conductivity $\sigma \sim T^{3/2}$
c	Goncz	[Gon65]	$1.13 \cdot 10^3$	1/2	$\varnothing_i = 1.3 \ldots 28$ mm $l_e = 6.3 \ldots 305$ mm $i = 13 \ldots 280$ A/mm²
d	Andreev	[AG68]	$1.9 \cdot 10^3$	1/3	A dependent on $\varnothing_i$ and pulse length, here: $\varnothing_i = 7$ mm, $\tau = 10^{-4}$ s
	Rakhimov	[RP71]	—	2/5	dissipation: heat conduction only
	Mavroyannakis	[Mav71]	—	2/5	conductivity $\sigma \sim T^{3/2}$ temperature $T \sim E$

in the parameter range indicated in the last column. Demenik et al. [DMSF68] and Rakhimov and Pismenny [RP71] estimate the exponent of (5.15) under opposite assumptions on the energy balance of the plasma column: the former neglect heat conduction and equate the electrical power to the blackbody radiation of the isothermal column while the latter take into account the heat conduction as the only dissipation process which cannot be accepted according to the estimates in section 3.1.1. Later Bakeev and Rovinski [BR70b] confirmed the measurements of Goncz with the exception that for gas pressures below $2 \cdot 10^4$ Pa the results become dependent on the filling pressure. Mavroyannakis [Mav71] points out the contradiction between the different expressions and proposes a formula with the power 2/5 based on the $T^{3/2}$-dependence of σ and a proportionality between T and E. On the whole, the graphic presentation of the different expressions in figure 5.1 yields considerable discrepancies of the resulting electrical field strengths which cannot be explained by experimental errors. The comparison with the dashed curves [Gü68] solves the problem: these curves were derived from field strength measurements under direct control of the

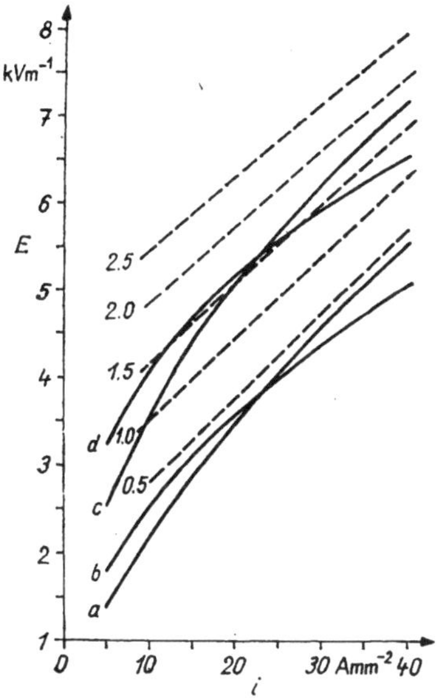

Fig. 5.1. Electrical characteristics of pulsed xenon discharges. Experimental curves of *a* MARSHAK [Mar78], *b* DEMENIK et al. [DMSF68], *c* GONCZ [Gon65], *d* ANDREEV and GAVRILOV [AG68]. Dashed lines: measurements of GÜNTHER [Gü68], Parameter: pressure in MPa

arc pressure and show, in contrast to the results of [AG68, DMSF68, Gon65, Mar78], a considerable dispersion of the curves in dependence on pressure which can be described by:

$$E = (Ap + B)\,i + (Cp + D) \tag{5.17}$$

with

$$A = -6.5 \cdot 10^{-12}\ \text{Pa}^{-1}\ \Omega\text{m}\,,$$

$$B = 1.00 \cdot 10^{-4}\ \Omega\text{m}\,,$$

$$C = 1.41 \cdot 10^{-3}\ \text{Pa}^{-1}\ \text{Vm}^{-1}\,,$$

$$D = 1.15 \cdot 10^{3}\ \text{Vm}^{-1}\,.$$

In this way, one can read easily that the experiments of MARSHAK and DEMENIK were made at arc pressures of about 0.3 MPa and those of GONCZ and ANDREEV at nearly 1.5 MPa. The reasons for these different arc pressures are probably different ratios of arc volume to dead space. As already mentioned, the empirical formulas listed in table 5.1, particularly the expression of GONCZ, were confirmed quantitatively in numerous measurements on commercial flashtubes. The change of arc pressure with the current density as well as its independence of the filling pressure is, besides the gas flows into the dead spaces, the result of many other mechanisms whose contribution varies over the parameter range. Therefore, a simple theoretical estimate similar to [DMSF68, Mav71, RP71] seems to be a questionable formalism rather than a physical explanation.

This result can be used for a nondestructive pressure diagnostics in closed discharge tubes under the condition that the arc fills the tube homogeneously. A simple measurement of field strength and current density is entered in figure 5.1, and the arc pressure can be read directly.

Table 5.2. Experimental studies on the electric properties of weakly nonideal plasmas

author	reference	device	element	temperature/ 10^4 K		pressure/MPa	
Bauder	[Bau76]	*A*	Ar	1 – 2.5	(α, β)*	0.1 – 20	(*a*)
Bauder et al.	[BDM73]	*A*	Ar	1 – 1.6	(α, β)*	0.1 – 20	(*a*)
Batenin and Minaev	[BM71]	*A*	Ar	1.2 – 1.5	(α, β)*	0.1 – 1	(*a*)
Goldbach et al.	[GNPP77, GNPP78]	*A*	Ar	1.1 – 1.3	(α)*	0.1 – 2	(*a*)
Kopainsky	[Ko71b]	*A*	Ar	1 – 2.5	(α, β)*	0.1 – 1	(*a*)
Glasser et al.	[GVC79]	*A*	Ar	1.1 – 1.3	(δ)*	1 – 2.5	(*a*)
Bakeev and Rovinski	[BR70b]	*C*	Ar	1.3 – 1.7	(δ)***	1 – 2	(*c*)
			Xe	1 – 1.4	(δ)***	0.3 – 1.8	(*c*)
Bakeev et al.	[BNRS71]	*C*	Kr	1.2 – 1.5	(δ)***	1.2 – 1.6	(*c*)
Ermokhin et al.	[EKKR71, EKKR78]	*F*	Cs	0.1 – 0.25		10^{-3} – 15	
Mitin et al.	[MKG75]	*D*	Ar	1.3 – 4.5	(δ)	10 – 95	(*b*)
Radtke and Günther	[RG76]	*B*	H	1.5 – 2.2	(γ)**	1	(*b*)
Günther et al.	[GLR83, GLRP77]	*B*	Ar	1.2 – 1.6	(γ)**	0.6 – 2	(*b*)
Günther and Radtke	[GR72]	*B*	Xe	1 – 1.2	(γ)**	1 – 2	(*b*)
Vuković and Popović	[VPP72]	*C*	Ar	1.3 – 1.9	(δ, β)***		
			Xe	1 – 1.6	(δ, β)***		
Hashiguchi et al.	[HI79a, HI79b, HSI77]	*E*	Ar	1.4 – 1.6	(γ)	2 – 8	(*b*)
			Xe	1.3	(γ)	8 – 20	(*b*)
Andreev and Gavrilov	[AG70b, AG70d]	*C*	Xe	1.1 – 1.9	(γ)****		(*d*)

Meaning of the symbols used in table 5.2:

Device:
- *A* stationary wall-stabilized cascade arc
- *B* pulsed wall-stabilized arc in cylindric quartz tubes
- *C* commercial flashtubes
- *D* pulsed high-pressure cascade arc
- *E* shock tube
- *F* other high-pressure facilities

Temperature determination:
- α from absolute line intensities
- β from the absolute continuum emission coefficient
- γ from Planck-Kirchhoff's law
- δ from equilibrium plasma composition calculations

Temperature profile:
- * from side-on observations of the spectral emission
- ** from end-on observations of the blackbody intensity
- *** assuming a homogeneous distribution
- **** using the distribution of discharge tubes with a rectangular cross-section [AG69b]

Pressure determination:
- *a* static measurement
- *b* using piezoresistive or piezoelectric probes
- *c* from the equation of state
- *d* from Dalton's law assuming constancy of heavy particle densities

5.3. Comparison of Published Electrical Conductivities

Information on transport properties of high-pressure ionized gases is necessary both for the design and optimization of discharges operating at elevated pressures and for checking the validity of theoretical approaches developed to the problem of nonideal charged particle interactions. Concerning the former intention the empirical expressions of the foregoing chapter are sufficient for many practical purposes. However, none of these techniques has resulted in any satisfactory conclusions about existing transport theories nor do they provide us with insights into dense plasma peculiarities. As regards this aspect of basic research, knowledge not only of electrical characteristics but also of thermodynamic state variables is necessary which requires measurements in well-conditioned plasmas using diagnostic methods whose interpretation is not or in a calculable manner influenced by nonideal charged particle interactions.

As can easily be realized from the examples dealt with in figure 1.2, weakly nonideal plasmas may favourably be generated by both stationary and pulsed wall-stabilized high-pressure arcs. The stationary wall-stabilized arc is widely used and has proved extremely profitable for measuring high-temperature gas transport properties. Bauder [BDM73, Bau76] demonstrated its flexibility by performing measurements in the pressure range up to 20 MPa with the same apparatus. Experiments with similar arc configurations at moderate pressure ranges have been reported by Batenin and Minaev [BM71], Goldbach et al. [GNPP77, GNPP78], Kopainsky [Ko71b] and Glasser et al. [GVC79] (see also table 5.2). Typically the arc was burnt in argon and the temperature achieved in the discharge varied between 12000 K and 15000 K corresponding to a pressure variation between 1 MPa and 3 MPa. The high pressure was maintained by means of a pressurized steel chamber containing the arc and could easily be controlled with a suitably designed gauge. For the determination of the arc temperature and its radial distribution the standard experimental techniques mentioned in the table were used. Most of the reported information on transport properties has been derived from the dc conductivity which was deduced from Ohm's

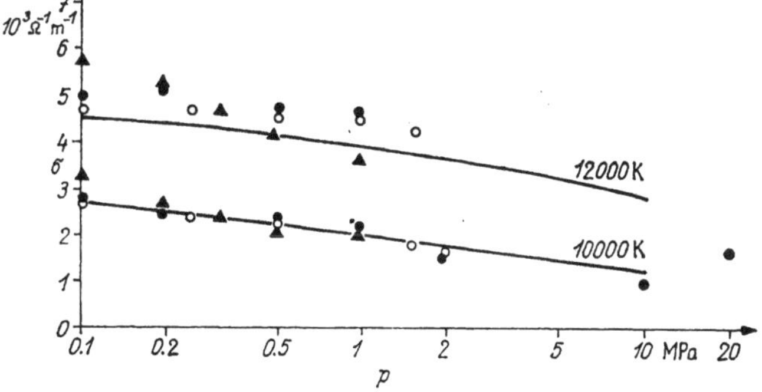

Fig. 5.2. Electrical conductivity of argon vs. pressure at fixed temperatures; ● Bauder et al. [BDM73]; ▲ Kopainsky [Ko71b]; ○ Goldbach et al. [GNPP78]; —— calculations according to Schirmer and Friedrich [SF58a, SF58b] including non-ideality corrections

law by measuring at least the electric field strength and temperature profiles as function of arc current. With additional assumptions concerning the dependence of σ on T as discussed in section 5.1., conductivities have been obtained in the form of functions $\sigma(T, p = \text{const})$ and $\sigma(p, T = \text{const})$. As an example, in figure 5.2 the latter is shown for argon at temperatures $T = 10000$ K and 12000 K. The experimental data presented by GOLDBACH et al. [GNPP77, GNPP78] agree very well with the results of BAUDER et al. [BDM73] and are at variance with those obtained by KOPAINSKY at 12000 K. Plotted also in figure 5.2 are values computed with the expressions given by SCHIRMER and FRIEDRICH [SF58a, SF58b]. In these calculations the corrections for the influence of the excited states of neutrals as well as for nonideality were taken into account by introducing the empirical screening correction $x(n_D) = r_s/D$ shown in figure 2.6 in the cross-section for charged particle interactions. The values of the measured conductivity at $T = 12000$ K were found to lie somewhat above theoretical predictions, whereas the experimental data at $T = 10000$ K are in satisfactory agreement with calculations.

To get an idea of the accuracy achieved in deriving electron transport coefficients from stationary arc experiments, in [BDM73] two possible sources of error were considered: first, the error involved in the measurement of the electric field strength which is mainly due to the leakage current from plate to plate through the cooling liquid, and secondly, the error due to uncertainties in the temperature profile. Concerning the former, some few per cent will not be exceeded at modest field strengths, but at higher field strengths present as the pressure is increased to values above 10 MPa errors of some ten per cent cannot be completely ruled out. The inaccuracies in the temperature profile are mainly due to the large changes in intensity with temperature near the wall. Moreover, in most of the experiments measurements were performed only to $r/R < 0.8$ (R being the arc radius) and extrapolations of varying severity were employed to carry out the integration over the cross-section necessary to solve the appropriate governing equations (concerning the details, see section 5.1.). In testing the sensitivity of the derived electrical conductivity to an erroneous temperature profile, a variation of about 10% was noted by BAUDER and co-workers [BDM73] for the range of temperature and pressure under consideration. This is in agreement with the estimates of other experimentalists and supports the conclusions reached by DEVOTO and MUKHERJEE [DM73] in re-analyzing a number of arc experiments at 0.1 MPa pressure. On the whole, the electrical conductivities derived from measurements on the stationary electric arc are reliable within 10 ... 15% including the uncertainties connected with determining the plasma state by using standard spectroscopic methods.

It is to be noted here that inconsistent conclusions have been derived from the above quoted investigations concerning the general trend observed in the nonideal behaviour of the electrical conductivity. So, for the Coulomb part of the conductivity GOLDBACH et al. [GNPP77, GNPP78] and BATENIN and MINAEV [BM71] reported values being below those from ideal plasma calculations. In addition, the latter not presented in figure 5.2 suggest much higher deviations from the ideal behaviour than indicated by GOLDBACH and co-workers which is apparently due to neglecting the effect of collisions with neutrals in the paper of BATENIN and MINAEV [BM71]. On the other hand, BAUDER et al. [BDM73, Bau76] concluded that in the condition of nonideal charged particle interaction the electrical conductivity is in excess of ideal-plasma calculations tending to confirm the trend predicted by the unified theory [GW67, It73, KA63, KAI63, Vi71, WW69]. Similar results have been reported recently by Glasser et al. [GVC79] who investigated the ac plasma conductivity taking advantage of the relationship which exists between it and the infrared continuum absorption [GC79]. In finding an explanation for the apparently contradictory results, two facts should be emphasized which have a direct influence on the reliability of the derived high-density corrections. First, the relatively low power input per arc length in the stationary electric arc sets a limit in the temperature range, and consequently, the values of degree of ionization are limited. The high electron density necessary for nonideal conditions is due to the high pressure, and the properties in these plasmas are dominated by electron-neutral interactions. In the high-pressure argon arc investigated by Bauder et al. [BDM73, Bau76], for example, n_D was of the order of unity, but the degree of ionization did not exceed 1%. Therefore, the Coulomb part of the electrical conductivity which must be selected from overall experimental data is quite sensitive to inaccuracies involved in this procedure, and it is difficult to relate possible deviations from the ideal behaviour to peculiarities in the charged particle interactions. Secondly, taking papers [GLR83, GNPP78] as a reference, a kind of threshold for non-ideality effects due to charge carrier interactions is observed at $n_D \approx 3$. In stationary electric arcs operating at sufficiently high degrees of ionization ($\alpha \geqq 10\%$) this corresponds almost to the lower limit of n_D-values where reliable measurements may be performed. Accordingly, the deviations from ideal-plasma calculations which may be expected under these conditions are small and possibly hidden by the uncertainties involved in the measurements.

In view of these circumstances, another class of experiments based on pulsed arc measurements was performed by Bakeev and Rovinski [BR70b], Bakeev et al. [BNRS71], Mitin et al. [MKG75], Radtke and Günther [GR72, RG76], Günther et al. [GLR83, GLRP77], Vuković and Popović [VPP72], and Andreev and Gavrilov [AG70b, AG70d]. As pointed out in section 3.2.5., pulsed arcs can reach rather high degrees of ionization even at elevated pressures which is certainly an advantage in studying weakly nonideal transport properties. Generally, the arc was burnt more than 1 ms at stable pressure and current which is sufficient time for the establishment of LTE conditions at higher electron concentrations as achieved in dense plasmas. The devices used for producing the pulsed arc as well as the temperature and pressure ranges covered by the experiments are summarized in table 5.2. Note that completely different diagnostics were used to determine the thermodynamic state variables which should be taken into consideration when the accuracy of published electrical conductivities has to be judged.

The most reliable techniques for the determination of the plasma state were employed by the authors of papers [GLR83, GLRP77, GR72, RG76] with respect to their sensitivity to dense plasma corrections (concerning the details, see sections 4.1.1. and 4.1.2.). Measurements of the electrical characteristics were performed for xenon [GR72], argon [GLR83, GLRP77] and hydrogen [RG76] and represented in the form of functions $\sigma(T, p = \text{const})$. The latter was derived with the help of the iterative optimization procedure described in section 5.1.2. taking as basic information the arc current, the electric field strength, and the radial temperature distribution derived from direct measurements. In addition, use of this method provides also information about the effective interaction of charged particles through the empirical screening

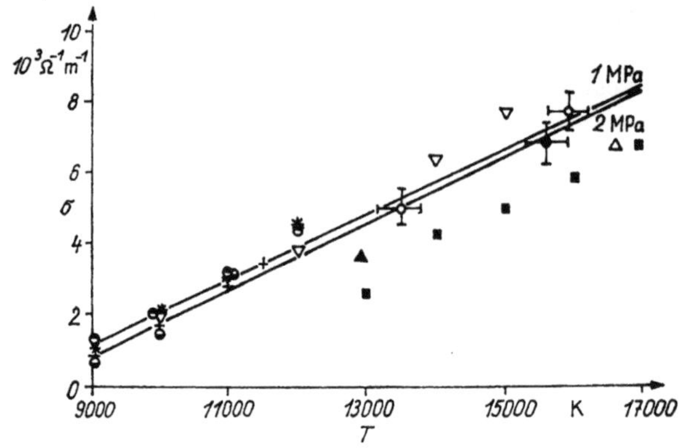

Fig. 5.3. Electrical conductivity of argon vs. temperature; ● GÜNTHER et al. [GLR GLRP77], $p = 1$ MPa; ○ as before, $p = 2$ MPa; ▲ BAKEEV and ROVINSKI [BR70b], $p = 1$ MPa; △ as before, $p = 2$ MPa; ■ VUKOVIĆ and POPOVIĆ [VPP72], $p \approx 1 \dots 2$ MPa; ◓ BAUDER et al. [BDM73], $p = 1$ MPa; ◒ as before, $p = 2$ MPa; ✳ GOLDBACH et al. [GNPP78], $p = 1$ MPa; + as before, $p = 2$ MPa; ▽ KOPAINSKY [Ko71b], $p = 1$ MPa; ——— calculations according to SCHIRMER and FRIEDRICH [SF58a, SF58b] including non-ideality corrections

function $x(n_D)$ as defined by (5.12). So, the values of x plotted in figure 2.6 refer to the experimental data [GLR83, GLRP77, GR72, RG76] and show the magnitude of non-ideality effects present in the n_D-range under consideration. The electrical conductivity values for argon, xenon and hydrogen obtained by the proposed procedure are shown in figures 5.3, 5.5, and 5.6, respectively, including the errors involved in the measurements. As regards the relative error in temperature determination δT, according to (4.6) it runs to 1 ... 2.5% corresponding to the variations of conditions in the experiment. The error in determining the electrical conductivity $\delta\sigma$ was estimated from the relative errors in arc current, field strength, and arc diameter and found not to exceed 6%. Note in this regard that the presented data refer to the axis conductivities in the centre of the discharge tube. In xenon arcs of high ionization level with characteristic flat profiles (see, e.g., the corresponding temperature distributions in figure 3.1) they differ only slightly from the average conductivities, but in plasmas

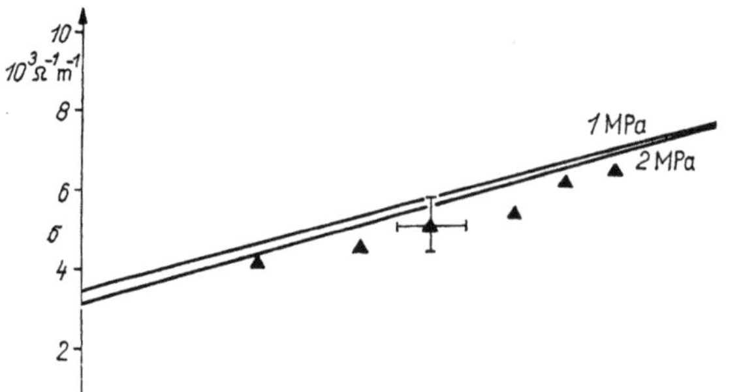

Fig. 5.4. Electrical conductivity of krypton vs. temperature; ▲ BAKEEV et al. [BNRS71], $p \approx 1.2 \dots 1.6$ MPa; ——— calculations according to SCHIRMER and FRIEDRICH [SF58a, SF58b] including non-ideality corrections

of the lighter rare gases or in the hydrogen are as investigated by RADTKE and GÜNTHER [RG76] a simple averaging of the current over the cross-section would result in much larger deviations since the plasma temperature, and consequently the electrical conductivity, depends on the radial coordinate rather sensitively.

Besides the results of GÜNTHER and RADTKE [GR72] and GÜNTHER et al. [GLR83, GLRP77], figures 5.3 and 5.5 contain also the measurements of BAKEEV and ROVINSKI [BR70b] and VUKOVIĆ and POPOVIĆ [VPP72] who deduced information on the plasma state using laser interferometric methods. Unfortunately, the data of the latter are represented only in the form of functions $\sigma(T, p_0 = \text{const})$ where p_0 is the initial pressure and no reliable estimate of the pressure during the discharge can be given. The results of BAKEEV and ROVINSKI, on the other hand, consist of data sets (σ, n_e, n_a) published later by ROVINSKI [Rov72a] in the form of tables so that both the arc temperature and the pressure could be derived from particle densities via Saha equation and equation of state. Taking into consideration possible uncertainties concerning the magnitude of high-density corrections that must be applied in performing the necessary equilibrium composition calculations combined with the errors involved in measuring particle number densities by laser interferometric methods, the derived values for temperature and pressure are expected to be accurate within 10% and 20%, respectively.

In comparing the data of BAKEEV and ROVINSKI [BR70b] and VUKOVIĆ and POPOVIĆ [VPP72] with those reported by GÜNTHER et al. [GLR83, GLRP77, GR72], the conductivities are seen to depart slightly from each other in the case of xenon and more strongly for argon. However, it is unlikely that errors of this magnitude are present

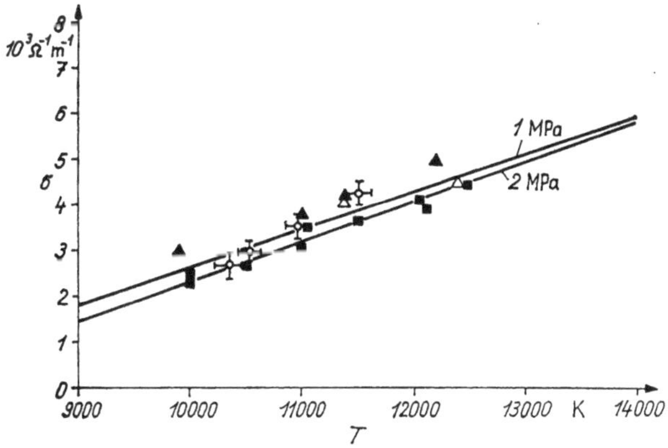

Fig. 5.5. Electrical conductivity of xenon vs. temperature; ○ GÜNTHER and RADTKE [GR72], $p = 2$ MPa; ▲ BAKEEV and ROVINSKI [BR70b], $p = 1$ MPa; △ as before, $p = 2$ MPa; ■ VUKOVIĆ and POPOVIĆ [VPP72], $p \approx 1 \ldots 2$ MPa; ——— calculations according to SCHIRMER and FRIEDRICH [SF58a, SF58b] including non-ideality corrections

in the experiment causing such a scatter of measured conductivities (as estimated by BAKEEV and ROVINSKI themselves [BR70b, Rov72a] their results are reliable within 15%). On the other hand, the spatial temperature distribution does not appear to have been taken into account in the analysis of experiments [BR70b, VPP72] which may result in considerable systematic errors exceeding those reported in the original publications. Evidently, the electrical conductivity at the centre of the arc is underestimated when a homogeneous temperature distribution is assumed and the amount of correction to find out the true value depends not only on the radial distribution actually observed in the experiment, but also on the trial function used in solving the appropriate governing equations. For xenon under typical arc conditions a modest correction of some few per cent is only required to account for the temperature decrease in the very narrow zone near the wall. However, in argon arc discharges of similar power dissipation the radial temperature profiles are curvatured and, therefore, much larger corrections must be applied to the average conductivity. Hence, the corrected conductivities are expected to lie substantially above those reported in the original analysis [BR70b, VPP72] which instructively illustrates the necessity of direct measurement of the spatial temperature distribution if specific quantities such as electrical conductivity have to be determined from integral electric measurements.

In addition to the results obtained with pulsed sources, figure 5.3 also shows the data derived from stationary arc experiments. The latter compare well with the calculations up to about 12000 K beyond which they slightly exceed both the theoretical curve and the data of Günther et al. [GLR83].

Experiments on krypton were performed by Bakeev et al. [BNRS71] and the results are shown in figure 5.4. The point indicator bars serve as an illustration of errors involved in the measurement. Within the estimated accuracy reasonable agreement with the theoretical predictions is to be noted.

Figure 5.6 is a comparison of calculated conductivities to the experimental values of Radtke and Günther [RG76] for hydrogen. It is obvious that under the conditions

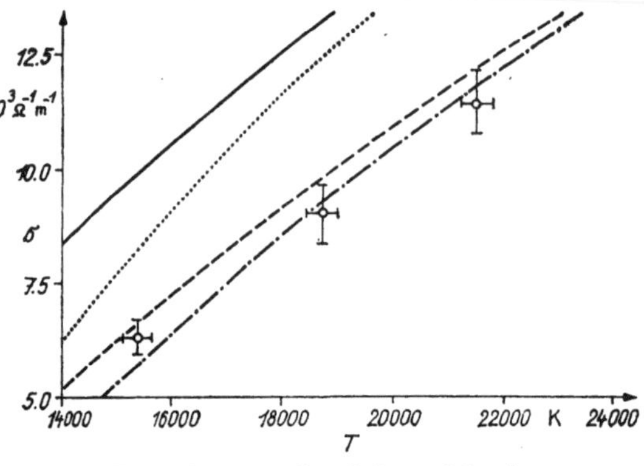

Fig. 5.6. Electrical conductivity of hydrogen vs. temperature; ○ Radtke and Günther [RG76], $p = 1$ MPa; calculations: fully ionized plasma result according to Spitzer [SH52] (———), according to Schirmer and Friedrich [SF58a, SF58b] with (— — —) and without (.........) non-ideality corrections, according to Khomkin [Kh78] (—·—·—)

of this experiment electron-atom collisions make a considerable contribution to the conductivity because of the large transport cross-section for hydrogen. The primary difference between experiment and ideal plasma calculations is associated with the specific features of the electron-ion interaction and can be removed through corrections to the classical binary collision theory. Also plotted is the theoretical result of Khomkin [Kh78] who used the idea of collision complexes. However, because of the larger electron-atom cross-section applied, KHOMKIN's conductivity values are lower than those obtained by the procedure described in section 2.3.2.

Table 5.2 also contains the data of some other experimental studies which have been carried out with special emphasis on the extent of non-ideality. The authors of papers [EKKR71, HI79a, HI79b, HSI77] have represented their results in the form of reduced conductivities and compared an experimentally determined screening distance with the Debye length. HASHIGUCHI and INUTAKE [HI79a] reported corrections similar to those proposed in figure 2.6 by the empirical screening function $x(n_D)$. On the other hand, the cesium data (ERMOKHIN et al. [EKKR71]) lie considerably above the predicted x-curve, but they seem to have qualitatively the same behaviour as the number of particles in the DEBYE sphere decreases. It seems that this discrepancy is associated with the difficulties encountered both in the diagnostics and interpretation of this particular cesium experiment. The results of MITIN et al. [MKG75] consist of data sets (p, I, E, n_0) covering in part the region of weakly non-ideal plasmas. However, it must be pointed out that the experimental errors are rather large resulting in a considerable scatter of measured conductivities ($\delta\sigma = 20-30\%$). None the less, their results do not contradict neither the experimental data of the other authors of table 5.2 nor the theoretical predictions mentioned in section 2.3.2.

6. Diagrams of Calculated Conductivities

Calculations of the electrical conductivity have been carried out for hydrogen, neon, argon, krypton, and xenon as well as for nitrogen and sulphurhexafluorid because of their relevance to many engineering problems. The results are plotted as functions of temperature and pressure in figures 6.1—6.7, respectively. For the calculation the method described by Schirmer and Friedrich [SF58a, SF58b] was used taking into account corrections for the influence of the excited states of neutrals as well as for non-ideality. The latter were introduced in the calculation through the empirical screening function $x(n_D)$ shown in figure 2.6. In the case of sulphurhexafluorid the interaction between electrons and neutral particles was described by a model of rigid spheres taking the corresponding sphere radii data from [Fri67]. The equilibrium plasma composition was computed in the usual way from the Saha equation using for the lowering of the ionization potential the expression recommended by Ecker and Kröll [EK66b].

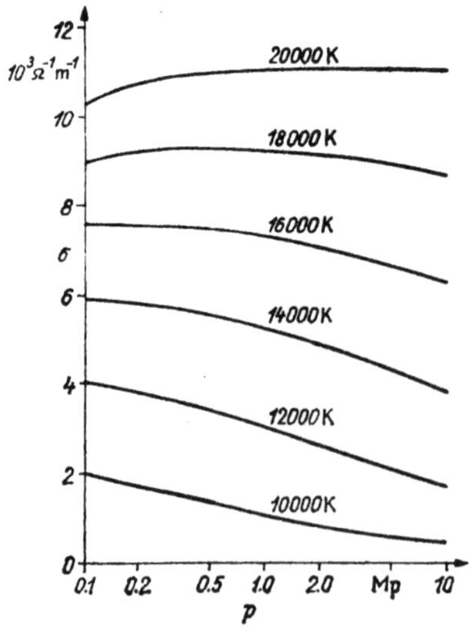

Fig. 6.1. Electrical conductivity of hydrogen

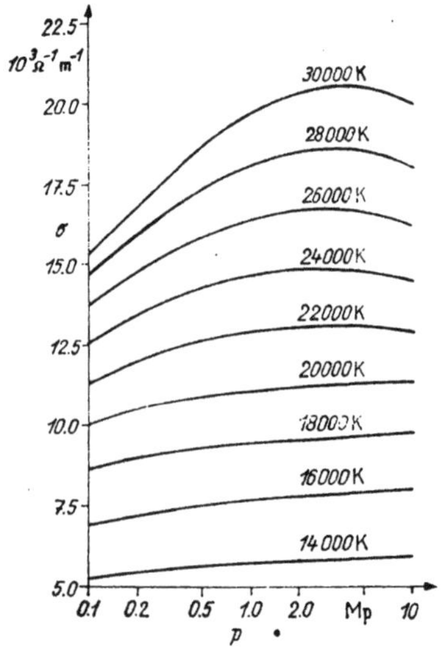

Fig. 6.2. Electrical conductivity of neon

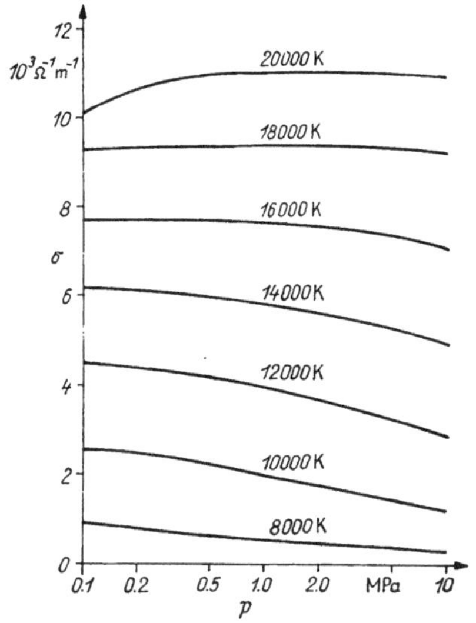

Fig. 6.3. Electrical conductivity of argon

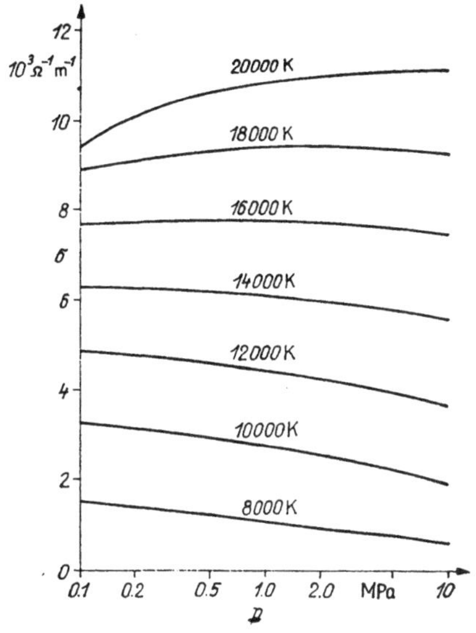

Fig. 6.4. Electrical conductivity of krypton

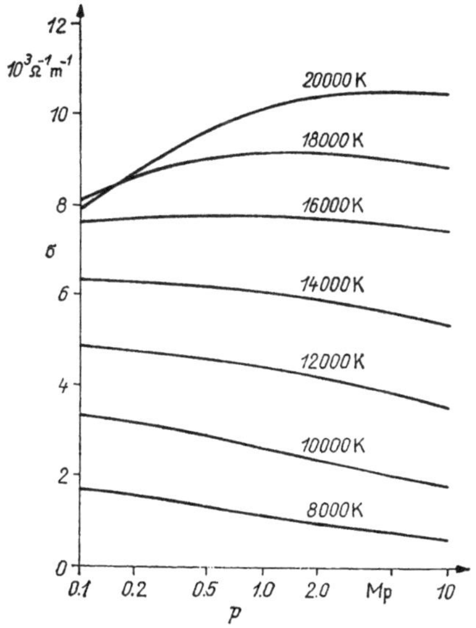

Fig. 6.5. Electrical conductivity of xenon

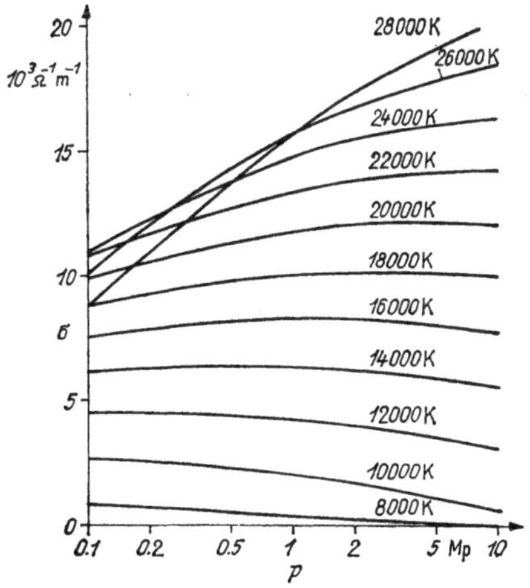

Fig. 6.6. Electrical conductivity of nitrogen

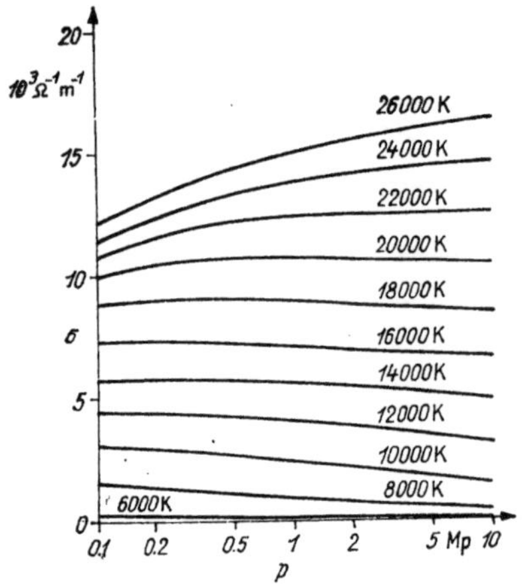

Fig. 6.7. Electrical conductivity of sulphurhexafluorid

7. Supplement

During the process of producing the book, there was a number of publications on the subject of nonideal plasmas both experimentally and theoretically. In this chapter, we want to call the reader's attention to these papers and to briefly point out the results in so far as the topic of the present work is concerned.

The progress in theory as well as in diagnostics and measurement of strongly nonideal plasmas has been reported in a series of review articles ([Fo82, AJ83, KKR83]). FORTOV [Fo82] actually reviewed the experimental schemes and designs of compressed and explosion-type nonideal plasmas including the diagnostic tools used. The results of experiments in the pressure range up to 10^6 MPa are reported, and the influence of the strong interparticle interaction on the thermodynamic, electric and optical properties is analyzed on the basis of model potentials accounting for screening due to the surrounding plasma.

The authors of paper [AJ83] deal with nonideal plasmas of metal vapours (especially Cs and Hg), and the parameters characteristic of the metal plasmas are $T \gtrapprox 1500$ K and $p \approx 10$ MPa. Under these conditions, the non-ideality is mainly due to the formation of electronic and ionic clusters leading to an essential change in the matter properties. For instance, the electrical conductivity can be higher by several orders of magnitude than the results of the standard estimates which makes nonideal plasmas of metal vapours a very attractive object of applications in the plasma technology (for example magnetohydrodynamic generators [BLYa82, Lik83]).

Measurements of the electrical conductivity have been reported by POISEL [Po83] (high-pressure cascade arc in Ar: $T = 10^4 \dots 1.5 \cdot 10^4$ K, $p = 0.2 \dots 20$ MPa), RUSCHE and JÄGER [RJ83] (pulsed capillary discharge in C: $T = 1.2 \cdot 10^4 \dots 3.5 \cdot 10^4$ K, $p = 100 \dots 300$ MPa), ROZANOV et al. [RKR83] (impulse capillary isobaric discharge in Li, Na, K and Cs: $T = 4 \cdot 10^3 \dots 1.4 \cdot 10^4$ K, $p = 10 \dots 100$ MPa) and LOMAKIN et al. [LLF83] (adiabatically compressed metal vapours of Cs and K: $T = 1.5 \cdot 10^3 \dots 4 \cdot 10^3$ K, $p = 2 \dots 10$ MPa).

The publications [Fi83, Uh83, LWBK83] describe experimental techniques to generate and investigate high-density plasmas, and the peculiarities in handling these sources are discussed (continuous optical discharge [Uh83]: $T = 1 \dots 2$ eV, $n_e = 10^{23} \dots 10^{24}$ m^{-3}; z-pinch [Fi83]: $T = 3 \dots 10$ eV, $n_e = 5 \cdot 10^{23} \dots 5 \cdot 10^{25}$ m^{-3}, laser ionization of sodium vapour [LWBK83]: $T \approx 0.2$ eV, $n_e \approx 10^{22}$ m^{-3}).

The paper [Hes83] gives an idea of the uncertainties involved when transforming the conductivity function $\sigma = \sigma(T, p)$ originally measured in an experiment into a representation $\sigma = \sigma(\mu)$ where μ is the plasma parameter (1.6). The latter form is in general use for theoretical considerations, but in extreme cases it can differ considerably depending on the treatment of particle interaction in equilibrium composition calculations (see also section 2.1.).

The problem of the electrical conductivity of fully ionized moderately nonideal plasmas has been attacked in several papers. WILHELM [Wil81a, Wil81b, Wil82], starting from the conductivity theory of ideal plasmas, has derived corrections for ν-body interactions ($\nu = 3, 4, \ldots$) of the electron from first principles of dimensional theory. This approach gives the strong power law dependence of the conductivity on the relevant dimensional parameters, whereas the dimensionless coefficients („COULOMB logarithm" for binary, triple etc. interactions) which are in general slowly varying functions have to be derived by means of a specific scattering model. Within a simple approximation for the scattering cross-section (DEBYE potential in first BORN approximation) WILHELM [Wil82] derived theoretical conductivity values which are roughly an order of magnitude smaller in the region $0.5 > n_D > 0.05$ than predicted by ideal plasma theory.

In papers [VK83, BKV83] the possible influence of plasma oscillations on the effective collision frequency was investigated. The effect of electron scattering by the microfields of plasma oscillations was estimated to give a contribution $\nu_p = \xi \cdot \omega_p$ (ω_p being the plasma frequency and $\xi = 0.016/n_D$) which is, for $\gamma \leqq 1$, comparable with that produced by the fields of individual particles. An interpolation formula for the electrical conductivity is derived and discussed in [VK83] which compares well with experimental data in the region $0.1 < \gamma < 1$.

Quantum-mechanical expressions for the transport coefficients of nonideal plasmas can be found in [KLM83, KBM83, HRRW84, BHS81, BRW82]. KRAEFT et al. [KLM83] determined the scattering phase shifts for the cut-off COULOMB potential, and as an example the electrical conductivity in the LORENTZ-gas approximation was derived. The results show that in the region of stronger non-ideality ($\mu > 0.5$) resonance scattering gives large contributions to the transport cross-section leading to minima in the conductivity function $\sigma(\mu)$. In the range $\mu < 0.5$, however, the resonance scattering is damped, and the classical limit result (2.59) is obtained.

In [HRRW84] the transport coefficients of partially ionized plasmas are expressed by equilibrium correlation functions in different approximations, and the effect of atoms was incorporated in the calculation via a polarization potential. The general dependence features of the transport coefficients and the effect of dynamical screening are discussed for temperatures of about 10^4 K and densities up to 10^{29} m^{-3}. The calculations suggest that the account of dynamical screening yields results which lie between the limiting cases of statical screening with electrons and ions (shielding length according to (1.4)) and statical screening by electrons only (shielding length according to (2.50)).

In papers [BHS81, BRW82] finally the conductivity of strongly coupled hydrogen plasmas was evaluated taking for the basis correlation functions as well as dynamic structure factors in different approximations. It is demonstrated that the LENARD-BALESCU result is obtained in the appropriate limits, and that a marked departure from the SPITZER regime occurs at larger coupling parameter $\Gamma = e^2(4\pi n_i/3)^{1/3}/kT$. In the weak-coupling limit corresponding to weakly nonideal plasmas, however, the effect of quantum corrections is small, and the numerical results are seen to agree well with the results of the standard estimates.

8. References

[ABD70] Andreev, S. I., O. G. Baikov and P. N. Dashuk (Андреев, С. И., О. Г. Байков и П. Н. Дашук), Teplofiz. vysok. temp. **8** (1970) 929

[ABDS69a] Andreev, S. I., O. G. Baikov, P. N. Dashuk, E. A. Sergeenkova and M. D. Yarysheva (Андреев, С. И., О. Г. Байков, П. Н. Дашук, Е. А. Сергеенкова и М. Д. Ярышева), J. tekhn. fiz. **39** (1969) 1025

[ABDS69b] Andreev, S. I., O. G. Baikov, P. N. Dashuk, E. A. Sergeenkova and M. D. Yarysheva (Андреев, С. И., О. Г. Байков, П. Н. Дашук, Е. А. Сергеенкова и М. Д. Ярышева), J. tekhn. fiz. **39** (1969) 1032

[ABSS77] Andreev, S. I., E. G. Borodachev, B. I. Sedov and A. K. Sinopalnikov (Андреев, С. И., Е. Г. Бородачев, Б. И. Седов и А. К. Синопальников), Teplofiz. vysok. temp. **15** (1977) 419

[AC70] Äschliman, D. P., and A. B. Cambel, Phys. Fluids **13** (1970) 2466

[AG68] Andreev, S. I., and V. E. Gavrilov (Андреев, С. И., и В. Е. Гаврилов), J. tekhn. fiz. **38** (1968) 1697

[AG69a] Andreev, S. I., and V. E. Gavrilov (Андреев, С. И., и В. Е. Гаврилов), Opt. i spektr. **26** (1969) 121

[AG69b] Andreev, S. I., and V. E. Gavrilov (Андреев, С. И., и В. Е. Гаврилов), Opt. i spektr. **26** (1969) 665

[AG70a] Andreev, S. I., and V. E. Gavrilov (Андреев,С. И., и В. Е. Гаврилов), J. priklad. spektr. **13** (1970) 988

[AG70b] Andreev, S. I., and V. E. Gavrilov (Андреев, С. И., и В. Е. Гаврилов), Вопросы физики низкотемп. плазмы. Nauka i tekhnika, Minsk 1970

[AG70c] Andreev, S. I., and V. E. Gavrilov (Андреев, С. И., и В. Е. Гаврилов), (Teplofiz. vysok. temp. **8** (1970) 203

[AG70d] Andreev, S. I., and V. E. Gavrilov (Андреев, С. И., и В. Е. Гаврилов), J. tekhn. fiz. **40** (1970) 1300

[AG71] Andreev, S. I., and V. E. Gavrilov, (Андреев, С. И., и В. Е. Гаврилов), J. priklad. spektr. **14** (1971) 310

[AJ83] Alekseev, V. A., and J. T. Jakubov, Physics Reports **96** (1983) 1

[AK77] Asinovski, E. I., and P. P. Kulik (Асиновский, Э. И., и П. П. Кулик), Inv. Lectures. XIII ICPIG, Berlin 1977, 1

[AKF73] Asinovski, E. I., N. A. Kozlov and V. V. Fomin (Асиновский, Е. И., Н. А. Козлов и В. В. Фомин), Teplofiz. vysok. temp. **11** (1973) 939

[Ande67] Anderson, J. E., Phys. Fluids **10** (1967) 894

[Ande68] Anderson, J. E., Phys. Fluids **11** (1968) 2060

[Andr67] Andreev, S. I. Андреев, С. И.), J. tekhn. fiz. **37** (1967) 1252

[AP70] Asinovski, E. I., and E. P. Pakhomov (Асиновский, Э. И., и Е. П. Пахомов), Teplofiz. vysok. temp. **8** (1970) 890

[At78] Attwood, D. T., IEEE J. Quant. Electr. **14** (1978) 909

[AVES67] Andreev, S. I., M. P. Vanyukov, A. A. Egorova and B. M. Sokolov (Андреев, С. И., М. П. Ванюков, А. А. Егорова и Б. М. Соколов), Opt. i spektr. **23** (1967) 784

[AW59] Adler, B., and T. Wainright, J. Chem. Phys. **31** (1959) 459
[Bar71] Barker, A. A., J. Chem. Phys. **55** (1971) 1751
[Bau76] Bauder, U., Appl. Phys. **9** (1976) 105
[Bau77] Bauder, U., J. de Phys. suppl. **38** (1977) C3-135
[BB79] Baker, E. A. M., and D. D. Burgess, J. Phys. **B12** (1979) 2097
[BDM73] Bauder, U., R. S. Devoto and D. Mukherjee, Phys. Fluids **16** (1973) 2143
[Bec74] Beck, G., Rev. Sci. Instr. **45** (1974) 318
[Beh71] Behringer, K., Z. Phys. **246** (1971) 333
[BGDF75] Bespalov, V. E., V. K. Gryaznov, A. N. Dremin and V. E. Fortov (Беспалов, В. Е., В. К. Грязнов, А. Н. Дремин и В. Е. Фортов) J. eksp. teor. fiz. **69** (1975) 2059
[BGLS77] Bushman, A. V., V. K. Gryaznov, B. N. Lomakin, V. A. Sechenov, V. E. Fortov and O. E. Shchekotov (Бушман, А. В., В. К. Грязинов, Б. Н. Ломакин, В. А. Сеченов, В. Е. Фортов и О. Е. Щекотов) Contr. Papers XIII ICPIG 1977, 587
[BHS81] Baus, M., J.-P. Hansen and L. Sjögren, Phys. Lett. **82A** (1981) 180
[BHU75] Baum, D., J. Hackmann and J. Uhlenbusch, Plasma Physics **17** (1975) 79
[BK75] Brünger, M., and M. Kock, Z. Naturforsch. **30a** (1975) 1560
[BKM80] Blümlein, J., W. D. Kraeft, and Th. Meyer, Ann. Phys. **37** (1980) 379
[BKN77] Biberman, L. M., V. Kovbasuk and A. Nedospasov (Биберман, Л. М., В. Ковбасук и А. Недоспасов), Inv. Lectures, XIII ICPIG, Berlin 1977, 35
[BKV83] Batenin, V. M., Yu. K. Kurilenkov and A. A. Valuev, Contr. Papers, XVI ICPIG, Düsseldorf 1983, 242
[BLSF75] Bushman, A. V., B. N. Lomakin, V. A. Sechenov, V. E. Fortov, O. E. Shchekotov and I. I. Sharipdzhanoye (Бушман, А. В., Б. Н. Ломакин, В. А. Сеченов, В. Е. Фортов, О. Е. Щекотов и И. И. Шарипджаное), J. eksp. teor. fiz. **69** (1975) 1624
[BLYa82] Biberman, L. M., A. A. Likalter and J. T. Jakubov (Биберман, Л. М., А. А. Ликальтер и И. Т. Якубов), Teplofiz. vysok. temp. **20** (1982) 565
[BM71] Batenin, V. M. and P. V. Minaev (Батенин, В. М. и П. В. Минаев), Teplofiz. vysok. temp. **9** (1971) 676
[BMM71] Bertinov, A. I., V. G. Manuilov and O. M. Mironov (Бертинов, А. И., В. Г. Мануйлов и О. М. Миронов), J. tekhn. fiz. **41** (1971) 1443
[BN67] Biberman, L. M., and G. E. Norman (Биберман, Л. М., и Г. Э. Норман), Usp. fiz. nauk **91** (1967) 193
[BNRS71] Bakeev, A. A., T. P. Narozhnaya, R. E. Rovinski and I. P. Shirokova (Бакеев, А. А., Т. П. Нарожная, Р. Е. Ровинский и И. П. Широкова), Teplofiz. vysok. temp. **9** (1971) 841
[BOS79] Bässler, P., H. U. Obbarius, and E. Schulz-Gulde, Physica **96c** (1979) 147
[BPMM77] Batenin, V. M., L. M. Pyatnitski, L. Ya. Margolin and P. V. Minaev (Батенин, В. М., Л. М. Пятницкий, Л. Я. Марголин и П. В. Минаев), Teplofiz. vysok. temp. **15** (1977) 239
[BPR67] Bues, I., H. J. Patt and J. Richter, Z. angew. Phys. **22** (1967) 345
[BR70a] Bakeev, A. A., and R. E. Rovinski (Бакеев, А. А., и Р. Е. Ровинский), Teplofiz. vysok. temp. **8** (1970) 207
[BR70b] Bakeev, A. A. and R. E. Rovinski (Бакеев, А. А., и Р. Е. Ровинский), Teplofiz. vysok. temp. **8** (1970) 1121
[BRS69] Bakeev, A. A., R. E. Rovinski nd I. P. Shirokova (Бакеев, А. А., Р. Е. Ровинский и И. П. Широкова), Opt. i. spektr. **17** (1969) 215
[BRS74] Bakeev, A. A., R. E. Rovinski and A. P. Sobolev (Бакеев, А. А., Р. Е. Ровинский и А. П. Соболев), Opt. i. spektr. **37** (1974) 1165
[BRW82] Boercker, D. B., F. J. Rogers and H. E. De Witt, Phys. Rev. **A25** (1982) 1623
[BS72] Bauder, U. H., and E. D. Stephens, Rev. Sci. Instr. **43** (1972) 1341
[BW65] Born, M., and E. Wolf, Principles of Optics. Pergamon Press, Oxford 1965
[CC52] Chapman, S., and T. G. Cowling, The Mathematical Theory of Non-uniform Gases. Cambridge University Press, New York 1952
[Chan43] Chandrasekhar, S., Rev. Mod. Phys. **15** (1943) 1

[Char68] Charnaya, F. A. (Ларная, Ф. А.), Opt. i spektr. **25** (1968) 24
[CHR75] Cilliers, W. A., J. D. Hey and J. P. S. Rash, J.Q.S.R.T. **15** (1975) 963
[CLL67] Cornevale, E. H., I. C. Lynnworth, and I. S. Larson, J. Chem. Phys. **46** (1967) 3040
[Co72] Conrads, H., Appl. Spectroscopy Rev. **6** (1972) 135
[CSL68] Church, C. H., R. G. Schlecht and I. Leberman, J.Q.S.R.T. 8 (1968) 403
[CSR50] Cohen, R. S., L. Spitzer and P. McRoutly, Phys. Rev. **80** (1950) 230
[DDES69] Dowling, J. A., J. Dovos, J. Eckerman, R. E. Schlier, J. Shumsky and P. Kisatsky, J. Appl. Optics **8** (1969) 1867
[Des63] Desloge, E. A., Americ. J. Phys. **31** (1963) 237
[Dev67] Devoto, R. S., Phys. Fluids **10** (1967) 354
[Dev73] Devoto, R. S., Phys. Fluids **16** (1973) 616
[DGRU77] Djordjević, D., K. Günther, R. Radtke and R. Ulbricht, Exp. Techn. Phys. **25** (1977) 433
[DH23] Debye, P., and E. Hückel, Physik. Z. **234** (1923) 305
[DH78] Dienemann, H., and H. Hess, 5th Intern. Conf. on Gas Discaharges, Liverpool 1978, IEE Conf. Pub. No. 165, 135
[DHTW77] Dienemann, H., H. Hess, W. Timm and M. Wirsig, Contr. Papers XIII ICPIG 1977, 591
[Di79] Dienemann, H., Thesis. University Greifswald 1979
[DM73] Devoto, R. S., and D. Mukherjee, J. Pl. Phys. **9** (1973) 65
[DMSF68] Demenik, I. V., V. E. Mnuskin, B. V. Skvortsov and V. B. Fedorov (Деменик, И. В., В. Е. Мнускин, Б. В. Скворцов и В. Б. Федоров), J. Tekhn. Fiz. **38** (1968) 1092
[DP73] Duschin, L. A., and O. S. Pawlitschenko, Plasmadiagnostik mit Lasern. Akademie-Verlag, Berlin 1973
[Dr56] Drawin, H. W., Z. Phys. **146** (1956) 295
[Dr68] Drawin, H. W., in: W. Lochte-Holtgreven, Plasma Diagnostics. North-Holland Publishing Company, Amsterdam 1968
[Dr69] Drawin, H. W., Z. Phys. **228** (1969) 99
[DRRZ77] Drobyshevski, E. M., B. I. Reznikov, S. I. Rozov and B. G. Zhukov (Дробышевский, Е. М., Б. И. Резников, С. И. Розов и Б. Г. Жуков), Contr. Papers XIII ICPIG 1977, 593
[DZ77] Dikhter, I. Ya., and V. A. Zeigarnik (Дихтер, И. Я., и В. А. Зейгарник), Teplofiz. vysok. **15** (1977) 196
[EFKK83] Ebeling, W., V. E. Fortov, Yu. L. Klimontovich, N. P. Kowalenko, W. D. Kraeft, Yu. P. Krasny, D. Kremp, P. P. Kulik, V. A. Riaby, G. Röpke, E. K. Rozanov and M. Schlanges, Transport Properties of Dense Plasmas, in: R. Rompe und M. Steenbeck, Ergebnisse der Plasmaphysik und der Gaselektronik, Vol. 6. Akademie-Verlag, Berlin 1983
[EK66a] Ecker, G., and W. Kröll, Z. Naturforsch. **21a** (1966) 2012
[EK66b] Ecker, G., and W. Kröll, Z. Naturforsch. **21a** (1966) 2023
[EK74] Ehrich, H., and H. J. Kusch, Z. Naturforsch. **29** (1974) 810
[EK79] Ermakov, V. V., and N. N. Kalitkin (Ермаков, В. В., и Н. Н. Калиткин), Fizika plazmy **5** (1979) 650
[EKK76] Ebeling, W., W. D. Kraeft and D. Kremp, Theory of Bound States and Ionization Equilibrium, in: R. Rompe und M. Steenbeck, Ergebnisse der Plasmaphysik und der Gaselektronik, Vol. 5. Akademie-Verlag, Berlin 1976
[EKKR71] Ermokhin, N. V., B. M. Kovalyov, P. P. Kulik and V. A. Ryaby (Ермохин, Н. В., Б. М. Ковалев, П. П. Кулик и В. А. Рябый), Teplofiz. vysok. temp. **9** (1971) 665
[EKKR78] Ermokhin, N. V., B. M. Kovalyov, P. P. Kulik and V. A. Ryab (Ермохин, Н. В., Б. М. Ковалев, П. П. Кулик и В. А. Рябый), J. Physique **39** (1978) 200
[Em67] Emmons, H. W., Phys. Fluids **10** (1967) 1125

[En12] ENSKOG, D., Ann. Phys. **38** (1921) 731

[ER79] EBELING, W., and G. RÖPKE, Ann. Phys. **36** (1979) 429

[Fi83] FINKEN, K. H., Fortschr. Phys. **31** (1983) 1

[FK72] FERZINGER, J. H., and H. G. KAPER, Mathematical Theory of Transport Processes in Gases. North-Holland Publishing Company, Amsterdam — London 1972

[FLDG76] FORTOV, V. E., A. A. LEONTEV, A. N. DREMIN and V. K. GRYAZNOV (Фортов, В. Е., А. А. Леонтьев, А. Н. Дремин и В. К. Грязнов), J. eksp. teor. fiz. **71** (1976) 216

[FM56] FINKELNBURG, W., and H. MAECKER, Handbuch der Physik 22. Springer Verlag, Berlin — Heidelberg — New York 1956

[Fo82] FORTOV, V. E. (Фортов, В. Е.), Usp. fiz. nauk **138** (1982) 361

[Fri67] FRIE, W., Z. Phys. **201** (1967) 269

[Frü65] FRÜNGEL, F., High speed pulse technology. Academic press, New York — London 1965

[Ga79] GAVRILOVA, T. V. (Гаврилова, Т. В.), J. tekhn. fiz. **49** (1979) 652

[GB66] GÜNTHER, K., and E. BELOV, Exp. Technik d. Phys. **14** (1966) 161

[GC77] GLASSER, J., and J. CHAPELLE, Contr. Papers, XIII ICPIG, Berlin 1977, 571

[GC79] GLASSER, J., and J. CHAPELLE, Plasma Physics **21** (1979) 477

[GCD75] GLASSER, J., J. CHAPELLE and D. DJORDJEVIČ, Contr. Papers, XII ICPIG, Eindhoven 1975, 379

[GE77] GUROVICH, V. T., and V. S. ENGELSHT (Гурович, в. т., и В. С. Енгельшт), J. eksp. teor. fiz. **72** (1977) 444

[GH75] GÜNTHER, K., and H. HESS, Proc. XII ICPIG, Eindhoven 1975, 139

[GIF77] GRYAZNOV, V. K., A. N. IVANOVA and V. E. FORTOV (Гряснов, В. К., А. Н. Йванова и В. Е. Фортов), Contr. Papers, XIII ICPIG, Berlin 1977, 589

[GKKM76] GLAZUNOV, G. P., V. P. KANTSEDAL, N. A. KOSIK, R. V. MITIN and A. P. SHEVSTUSEV (Глазунов, Г. П., В. Т. Канцедал, Н. А. Косик, Р. В. Митин и А. П. Шевстусев), Teplofiz. vysok. temp. **14** (1976) 1175

[GLR83] GÜNTHER, K., S. LANG and R. RADTKE, J. Phys. **D16** (1983) 1235

[GLRP77] GÜNTHER, K., S. LANG, R. RADTKE, S. S. POPOVIĆ and M. M. POPOVIĆ, Contr. Papers, XIII ICPIG, Berlin 1977, 577

[GNP77] GOLDBACH, C., G. NOLLEZ and P. PLOMDEUR, J. Phys. B **10** (1977) 1181

[GNPP77] GOLDBACH, C., G. NOLLEZ, M. M. POPOVIĆ and S. S. POPOVIĆ, Contr. Papers, XIII ICPIG, Berlin 1977, 575

[GNPP78] GOLDBACH, C., G. NOLLEZ, S. S. POPOVIĆ and M. M. POPOVIĆ, Z. Naturforsch. **39a** (1978) 11

[Gon65] GONCZ, J. H., J. Appl. Phys. **36** (1965) 742

[GPPR76] GÜNTHER, K., M. M. POPOVIĆ, S. S. POPOVIĆ and R. RADTKE, J. Phys. **D9** (1976) 1139

[GR72] GÜNTHER, K., and R. RADTKE, Beitr. Plasmaphys. **12** (1972) 63

[GR73] GÜNTHER, K., and R. RADTKE, Contr. Papers, XI ICPIG, Prague 1973, 446

[GR75] GÜNTHER, K., and R. RADTKE, J. Phys. **E8** (1975) 371

[Gra61] GRAD, H., Proceedings, V ICPIG, München 1961, 1630

[Gri63] GRIEM, H. R., Phys. Rev. **131** (1963) 1170

[Gri64] GRIEM, H. R., Plasma Spectroscopy. McGraw-Hill, New York 1964

[Gri74] GRIEM, H. R., Spectral Line Broadening by Plasmas. Academic Press, New York 1974

[Gri78] GRIEM, H. R., Phys. Rev. **A17** (1978) 214

[Gu68] GÜNTHER, K., Beitr. Plasmaphys. 8 (1968) 383

[Gu70a] GÜNTHER, K., Thesis, Akademie der Wissenschaften der DDR, Berlin 1970

[Gu70b] GÜNTHER, K., Beitr. Plasmaphys. **10** (1970) 469

[Gu73] GÜNTHER, K., Opt. i spektr. **34** (1973) 1070

[Gund70] GÜNDEL, H., Beitr. Plasmaphys. **10** (1970) 455

[Gund71] GÜNDEL, H., Beitr. Plasmaphys. **11** (1971) 1

[Gur79] GUREVICH, I. M. (Гуревич, И. М.), Teplofiz. vysok. temp. **17** (1979) 906
[Gus73] GUSINOW, M. A., J. Appl. Phys. **44** (1973) 4567
[GVC78] GLASSER, J., R. VILLADROSA and J. CHAPELLE, J. Phys. **D11** (1978) 1703
[GVC79] GLASSER, J., R. VILLADROSA and J. CHAPELLE, Contr. Papers, XIV ICPIG, Grenoble 1979, 687
[GW67] GOULD, H. A., and H. E. DEWITT, Phys. Rev. **155** (1967) 68
[GW77] GRÜTZMACHER, K., and B. WENDE, Phys. Rev. **A16** (1977) 243
[GW78] GRÜTZMACHER, K., and B. WENDE, Phys. Rev. **A18** (1978) 2140
[GZKM70] GENERALOV, N. A., V. P. ZIMAKOV, G. I. KOZLOV, V. A. MASYUKOV and YU. P. RAIZER (Генералов, Н. А., В.П. Зимаков, Г. И. Козлов, В. А. Масюков и Ю. П. Райзер), Pisma J. eksp. teor. fiz. **11** (1970) 447
[GZKM71] GENERALOV, N. A., V. P. ZIMAKOV, G. I. KOZLOV, V. A. MASYUKOV and YU. P. RAIZER (Генералов, Н. А., В. П. Зимаков, Г. И. Козлов, В. А. Масюков и Ю. П. Райзер), J. eksp. teor. fiz. **61** (1971) 1134
[HCB54] HIRSCHFELDER, J. O., CH. F. CURTISS and R. B. BIRD, Molecular theory of gases and liquids, J. Wiley Sons, Inc., New York 1954
[He68a] HERMANN, W., Z. Phys. **216** (1968) 33
[He68b] HERMANSDORFER, H., in: W. LOCHT-HOLTGREVEN, Plasma Diagnostics. North-Holland Publishing Company, Amsterdam 1968
[Hes83] HESS, H., Beitr. Plasmaphys. **23** (1983) 513
[HG77] HESS, H., and K. GÜNTHER, J. Phys. **D**, Appl. Phys. **10** (1977) L223
[HI79a] HASHIGUCHI, S., and M. INUTAKE, Contr. Papers, XIV ICPIG, Grenoble 1979, 681
[HI79b] HASHIGUCHI, S., and M. INUTAKE, Report IPPJ-376, University Nagoya 1979
[HK67] HEARNE, K. R., and N. KONJEVIĆ, Z. Phys. **204** (1967) 443
[HM67] HERMANN, W., and A. MONTERDE-GARCIA, Z. Phys. **205** (1967) 3/3
[HM68] HOCHSTIM, A. R., and G. A. MASSEL, Kinetic Processes in Gases and Plasmas. Academic Press, New York—London 1969
[HMMS72] HAHN, H. S., E. A. MASON, E. J. MILLER, and S. J. SEDLER, J. Pl. Phys. **7** (1972) 285
[HMS71] HAHN, H. S., E. A. MASON, and F. J. SMITH, Phys. Fluids **14** (1971) 278
[HMU72] HACKMANN, J., H. MICHAEL and J. UHLENBUSCH, Z. Phys. **250** (1972) 207
[Hon64] HONDA, N., J. Phys. Soc. Jap. **19** (1964) 1201
[Hoo72] HOOK, W. R., Dishington and Hilberg, IEEE Transactions on Electron Devices **ED-19** (1972) 308
[HRRW84] HÖHNE, F. E., R. REDMER, G. RÖPKE and H. WEGENER, Physica **A** (in press)
[HS63] HERMAN, F., and S. SKILLMAN, Atomic Structure Calculation. Prentice-Hall, Inc., Englewood Cliffs (N. J.) 1963
[HS70] HERMANN, W., and E. SCHADE, Z. Phys. **233** (1970) 333
[HSI77] HASHIGUCHI, S., S. SEKIMOTO and M. INUTAKE, 11th Int. Symp. on Shock Tubes and Waves, Seattle 1977, 622
[HZ82] HÖHNE, F. E., and R. ZIMMERMANN, J. Phys. **B 15** (1982) 2551
[IMDF76] IVANOV, YU. V., V. B. MINTSEV, V. E. FORTOV and A. N. DREMIN (Иванов, Ю. В., В. Б. Минцев, В. Е. Фортов и А. Н. Дремин), J. eksp. teor. fiz. **71** (1976) 216
[It63] ITIKAWA, Y., J. Phys. Soc. Jap. **18** (1963) 1499
[KA63] KIHARA, T., and O. AONO, J. Phys. Soc. Jap. **18** (1963) 837
[KAI63] KIHARA, T., O. AONO and Y. ITIKAWA, J. Phys. Soc. Jap. **18** (1963) 1043
[Kal70] KALITKIN, N. N. (Калиткин, Н. Н.), Teplofiz. vysok. temp. **8** (1970) 1121
[Kap69] KAPITSA, P. L. (Капица, П. Л.), J. eksp. teor. fiz. **57** (1969) 1801
[KBM83] KRAEFT, W. D., J. BLÜMLEIN and T. MEYER, Beitr. Plasmaphys. **23** (1983) 9
[Ke38] KENNARD, E. H., Kinetic Theory of Gases. McGraw-Hill Book Company, Inc., New York 1938
[KF76a] KOZLOV, N. A., and V. V. FOMIN (Козлов, Н. А., и В. В. Фомин) Teplofiz. vysok. temp. **14** (1976) 457
[KF76b] KURILENKOV, YU. K., and V. S. FILINOV (Куриленков, Ю. К., и В. С. Филинов), Teplofiz. vysok. temp. **14** (1976) 886

[Kh74] Khomkin, A. L. (Хомкин, А. Л.), Teplofiz. vysok. temp. **12** (1974) 879
[Kh78] Khomkin, A. L. (Хомкин, А. Л.), Teplofiz. vysok. temp. **16** (1978) 37
[Ki64] Kihara, T., J. Phys. Soc. Japan **19** (1964) 108
[KK74a] Kunavin, A. T., and Yu. S. Korshunov (Кунавин, А. Т., и Ю. С. Коршунов), Teplofiz. vysok. temp. **12** (1974) 1302
[KK74b) Konkov, A. A., and S. G. Kulagin (Конков, А. А., и С. Г. Кулагин), Teplofiz. vysok. temp. **12** (1974) 497
[KK78] Kobtsev, G. A., and Yu. K. Kurilenkov (Кобцев, Г. А., и Ю. К. Куриленков), Teplofiz. vysok. temp. **16** (1978) 458
[KKM74a] Kozlov, G. I., V. A. Kuznetsov and V. A. Masyukov (Козлов, Г. И., В. А. Кузнецов и В. А. Масюков), J. eksp. teor. fiz. **66** (1974) 954
[KKM74b] Kozlov, G. I., V. A. Kuznetsov and V. A. Masyukov (Козлов, Г. И., В. А· Кузнецов и В. А. Масюков), Opt. i. spektr. **37** (1974) 1049
[KKM75] Kozlov, G. I., V. A. Kuznetsov and V. A. Masyukov (Козлов, Г. И., В. А. Кузнецов и В. А. Масюков), Fizika plazmy **1** (1975) 830
[KKN71] Krasnikov, Yu. G., P. P. Kulik and G. E. Norman (Красников, Ю. Г., П. П. Кулик и Г. Э. Норман), Inv. Papers, X ICPIG, Oxford 1971, 405
[KKN77] Kobtsev, G. A., Yu. K. Kurilenkov and G. E. Norman (Кобцев, Г. А., Ю. К. Куриленков и Г. Э. Норман), Teplofiz. vysok. temp. **15** (1977) 193
[KKPR77] Kovalyov, B. M., P. P. Kulik, A. Yu. Popov and V. A. Ryaby (Ковалев, Б. М., П. П. Кулик, А. Ю. Попов и В. А. Рябый), Contr. Papers, XIII ICPIG, Berlin 1977, 597
[KKR72] Kovalyov, B. M., P. P. Kulik and V. A. Ryaby (Ковалев, Б. М., П. П. Кулик и В. А. Рябый), Engin. fiz. J. **22** (1972) 92
[KKR83] Kraeft, W. D., D. Kremp and G. Röpke, Inv. Papers, XVI ICPIG, Düsseldorf **1983**, 24
[Kle68] Klein, L., Appl. Opt. **7** (1968) 677
[Kli75] Klimontovich, Yu. L. (Климонтович, Ю. Л.), Кинетическая теория неидеального газа и неидеальной плазмы, Nauka, Moscow 1975
[KLM83] Kraeft, W. D., M. Luft and A. A. Mihajlov, Physica **120A** (1983) 263
[KM67] Kusch, H. J., and E. R. Mewes, Z. Naturforsch. **22a** (1967) 676
[KM78] Kurilenkov, Yu. K., and P. V. Minaev (Куриленков, Ю. К., и П. В. Минаев), J. eksp. teor. fiz. **74** (1978) 563
[KMG73] Kantsedal, V. P., R. V. Mitin and G. P. Glazunov (Канцедал, В. П., Р. В. Митин и Г. П. Глазунов), Contr. Papers, XI ICPIG, Prague 1973, 237
[KMRT73] Kulik, P. P., V. M. Melnikov, V. A. Ryaby, and M. A. Titov (Кулик, П. П., В. М. Мельников, В. А. Рябый и М. А. Титов), Contr. Papers, XI ICPIG, Prague 1973, 423
[KN73] Kaklyugin, A. S., and G. E. Norman (Каклюгин, А. С., и Г. Э. Норман), Teplofiz. vysok. temp. **11** (1973) 238
[Ko71a] Kopainsky, J., Z. Phys. **248** (1971) 405
[Ko71b] Kopainsky, J., Z. Phys. **248** (1971) 417
[KR74] Kozlov, G. I., and D. I. Roitenburg (Козлов, Г. И., и Д. И. Роитенбург), Opt. i spektr. **36** (1974) 850
[KRR77] Kulik, P. P., E. K. Rosanov and V. A. Ryaby (Кулик, П. П., Е. К. Росанов и В. А. Рябый), Teplofiz. vysok. temp. **15** (1977) 415
[Ku72] Kulik, P. P. (Кулик, П. П.), AIAA Paper No. 72-414 (1972)
[LF71] Lomakin, B. N., and V. E. Fortov (Ломакин, Б. Н., и В. Е. Фортов), Teplofiz. vysok. temp. **9** (1971) 1291
[LF72] Lomakin, B. N., and V. E. Fortov (Ломакин, Б. Н., и В. Е. Фортов), J. eksp. teor. fiz. **63** (1972) 92
[LFS70] Lomakin, B. N., V. E. Fortov and O. E. Shchekotov (Ломакин, Б. Н., В. Е. Фортов и О. Е. Щекотов), Teplofiz. vysok. temp. 8 (1970) 154
[Lib59] Liboff, R. L., Phys. Fluids 2 (1959) 40
[Lik83] Likalter, A. A. (Ликалътер, А. А.), Teplofiz. vysok. temp. **21** (1983) 249
[Lit73] Litvinov, I. I. (Литвинов, И. И.), Teplofiz. vysok. temp. **4** (1973) 695
[Lit75] Litvinov, I. I. (Литвинов, И. И.), Teplofiz. vysok. temp. **13** (1975) 1100

[LL74] LANDAU, L. D., and E. M. LIFSCHITZ, Elektrodynamik der Kontinua, in: Lehrbuch der Theoretischen Physik, Vol. VIII, 3rd Ed. Akademie-Verlag, Berlin 1974

[LLF76] LITVINOV, I. I., E. D. LYUMKIS and S. S. FILIPPOV (Литвинов, И. И., Е. Д. Люмкис и С. С. Филиппов), Probl. mat. teor. fiz. **1** (1976) 27

[LLF83] LOMAKIN, B. N., A. D. LOPATIN and V. E. FORTOV, Contr. Papers, XVI ICPIG, Düsseldorf 1983, 234

[Loc68] LOCHTE-HOLTGREVEN, W., Plasma diagnostics. North-Holland Publ. Comp., Amsterdam 1968

[Low70] LOWKE, J. J., J. Appl. Phys. **1** (1970) 2588

[Low74] LOWKE, J. J., J.Q.S.R.T. **14** (1974) 111

[Lu73] LUTHJENS, L. H., Rev. Sci. Instr. **44** (1973) 1661

[LWBK83] LANDEN, O. L., R. J. WINFIELD, D. D. BURGESS, J. D. KILKENNY and R. W. LEE, Contr. Papers, XVI ICPIG, Düsseldorf 1983, 258

[Mar78] MARSHAK, I. S. (Маршак, И. С.), Импульсные источники света. Energiya, Moscow 1978

[Mav71] MAVROYANNAKIS, E. G., IEEE Transactions on Electron Devices **ED-18** (1971) 115

[MKG75] MITIN, R. V., V. P. KANTSEDAL and G. P. GLAZUNOV (Митин, Р. В., В. П. Канцедал и Г. П. Глазунов), Teplofiz. vysok. temp. **13** (1975) 706

[MMS67] MASON, E. A., R. J. MUNN and F. J. SMITH, Phys. Fluids **10** (1967) 1827

[MN77] MERCIER, J., and C. R. NEUFELD, J. Appl. Phys. **48** (1977) 4498

[MP69] MITIN, R. V., and K. K. PRYADKIN, (Митин, Р. В., и К. К. Прядкин), Magnitn. lobushki **4** (1969) 84

[MPS52] MAECKER, H., T. PETERS and H. SCHENK, Z. Phys. **140** (1952) 119

[MPZ70] MITIN, R. V., K. K. PRYADKIN and A. V. ZVYAGINTSEV (Митин, Р. В., К. К. Прядкин и А. В. Звягинцев), Teplofiz. vysok. temp. **8** (1970) 1142

[MR82] MEISTER, C.-V., and G. RÖPKE, Ann. Phys. **39** (1982) 133

[MRV70] MORRIS, J. C., R. P. RUDIS and J. M. VOS, Phys. Fluids **13** (1970) 608

[MS66] MAGDEBURG, H., and U. SCHLEY, Z. angew. Phys. **20** (1966) 465

[MS67] MAECKER, H., and S. STEINBERGER, Z. angew. Phys. **23** (1967) 456

[MW76] MEINERS, D., and C. O. WEISS, J.Q.S.R.T. **16** (1976) 273

[MZP73] MITIN, R. V., A. V. ZVYAGINTSEV and K. K. PRYADKIN (Митин, Р. В., А. В. Звягинцев и К. К. Прядкин), Teplofiz. vysok. temp. **11** (1973) 493

[Ne75] NEUMANN, W., Spectroscopic Methods of Plasma Diagnostics, in: R. ROMPE and M. STEENBECK, Progress in Plasmas and Gas Electronics, Vol. 1. Akademie-Verlag, Berlin 1975

[No79] NORMAN, G. E. (Норман, Г. Э.), Teplofiz. vysok. temp. **17** (1979) 453

[NP72] NIKOLAEVSKI, V. G., and V. D. PISMENNY (Николаевский, В. Г., и В. Д. Письменный) J. tekhn. fiz. **42** (1972) 364

[NS68] NORMAN, G. E., and A. N. STAROSTIN (Норман, Г. Э., и А. Н. Старостин), Teplofiz. vysok. temp. **6** (1968) 410

[NV79] NORMAN, G. E., and A. A. VALUEV, Plasma Physics **21** (1979) 531

[OPS77] OGURTSOVA, N. N., I. V. PODMOSHENSKI and V. L. SMIRNOV (Огурцова, Н. Н., И. В. Подмошенский и В. Л. Смирнов), Teplofiz. vysok. temp. **15** (1977) 456

[OPZ69] OGURTSOVA, N. N., I. V. PODMOSHENSKI and V. M. ZHELEMINA (Огурцова, Н. Н., И. В. Подмошенский и В. М. Желемина), Teplofiz. vysok. temp. **7** (1969) 213

[Os60] OSTER, L., Rev. Modern Phys. **32** (1960) 141

[PBZM78] PYATNITSKI, L. N., V. M. BATENIN, V. S. ZHIVOPISTSEV, L. YA. MARGOLIN and P. V. MINAEV (Пятницкий, Л. Н., В. М. Батенин, В. С. Живописцев, Л. Я. Марголин и П. В. Минаев), Teplofiz. vysok. temp. **16** (1978) 202

[Pe66] PERLMANN, D. E., Rev. Sci. Instr. **37** (1966) 340

[PK76] POPOVIĆ, P., and N. KONJEVIĆ, Z. Naturforsch. **31a** (1976) 1042

[Pl69] PLANTIKOW, U., Z. Phys. **227** (1969) 271

[PMK69] PRYADKIN, K. K., R. V. MITIN and N. N. KLIMOV (Прядкин, К. К., Р. В. Митин и Н. Н. Климов), J. priklad. spektr. **10** (1969) 721
[Po83] POISEL, H., Contr. Papers, XVI ICPIG, Düsseldorf 1983, 60
[Pod75] PODLUBNY, L. N. (Подлубный, Л. Н.), Teplofiz. vysok. temp. **13** (1975) 1081
[Pop77] POPOVIĆ, S, S., Thesis, University Belgrade (1977)
[PP79] PROSKUROVSKI, D. I., and V. F. PUCHKAREV (Проскуровский, Д. И. и В. Ф. Пучкарев), J. tekhn. fiz. **49** (1979) 2611
[PPV74] POPOVIĆ, M. M., S. S. POPOVIĆ and S. M. VUKOVIĆ, Fizika **6** (1974) 29
[Pr77] PRESTON, R. C., J. Phys. **10** (1977) 2381
[PR80] PODLUBNY, L. I., and V. S. RAGOVSKI (Подлубный, Л. И., и В. С. Раговский), Fizika plazmy **6** (1980) 1339
[PS70] PLANTIKOW, U., and S. STEINBERGER, Z. Phys. **231** (1970) 109
[PT65] PRESSLEY, R. J., and H. Y. S. TANG, Rev. Sci. Instr. **36** (1965) 1381
[Py75] PYATNITSKI, L. N. (Пятницкий, Л. Н.), Teplofiz. vysok. temp. **13** (1975) 1288
[Ra74] RADTKE, R., Thesis, University Berlin 1974
[RDJ57] ROSENBLUTH, M. N., W. M. MCDONALD and D. L. JUDD, Phys. Rev. **107** (1957) 1
[REK80] RÖPKE, G., W. EBELING and W. D. KRAEFT, Physica **101A** (1980) 243
[RG] RADTKE, R., and K. GÜNTHER, to be published
[RG73] RADTKE, R., and K. GÜNTHER, Contr. Papers, XI ICPIG, Prague 1973, 274
[RG75] RADTKE, R., and K. GÜNTHER, Beitr. Plasmaphys. **15** (1975) 300
[RG76] RADTKE, R., and K. GÜNTHER, J. Phys. **D9** (1976) 1131
[RGU80] RADTKE, R., K. GÜNTHER and R. ULBRICHT, J. Phys. **D13** (1980) 1
[RJ83] RUSCHE, D., and H. JÄGER, Contr. Papers, XVI ICPIG, Düsseldorf 1983, 228
[RKR83] ROZANOV, E. K., P. P. KULIK and V. A. RYABY, Contr. Papers, XVI ICPIG, Düsseldorf 1983, 232
[RMKK79] RÖPKE, G., C.-V. MEISTER, K. KOLLMORGEN and W. D. KRAEFT, Ann. Phys. **36** (1979) 377
[Rob67] ROBINSON, J. M., J. Appl. Phys. **38** (1967) 210
[Rog80] ROGERS, F. J., Analytic electron-ion effective potentials for $z \leqq 55$, UCRL-84868 (1980)
[Ros75] ROSZMAN, L. J., Phys. Rev. Lett. **34** (1975) 785
[Röp82] RÖPKE, G., Ann. Phys. **39** (1982) 35
[Rov72a] ROVINSKI, R. E. (Ровинский, Р. Е.), Teplofiz. vysok. temp. **10** (1972) 1
[Rov72b] ROVINSKI, R. E. (Ровинский, Р. Е.), J. tekhn. fiz. **12** (1972) 1869
[Rov75] ROVINSKI, R. E. (Ровинский, Р. Е.), J. teckhn. fiz. **45** (1975) 1782
[Rov77] ROVINSKI, R. E. (Ровинский, Р. Е.), Teplofiz. vysok. temp. **15** (1977) 233
[RP71] RAKHIMOV, A. I., and V. D. PISMENNY (Рахимов, А. И., и В. Д. Письменный), Doklad. Akad. Nauk SSSR **196** (1971) 812
[RSL78] RAMAKRISHNAN, S., A. D. STOKES and J. J. LOWKE, J. Phys. **D 11** (1978) 2267
[RV76] RUZDIAK, V., and V. VUJNOVIĆ, Astron. and Astrophys. **54** (1976) 751
[RWB81] ROGERS, F. J., H. E. DEWITT and D. B. BOERCKER, Phys. Lett. **82A** (1981) 331
[Schi60] SCHIRMER, H., Thesis, University Berlin 1960
[Schl68] SCHLÜTER, D., Z. Phys. **210** (1968) 80
[Se77a] SEIDEL, J., Z. Naturforsch. **A 32** (1977) 1195
[Se77b] SEIDEL, J., Z. Naturforsch. **A 32** (1977) 1207
[SF58a] SCHIRMER, H., and J. FRIEDRICH, Z. Phys. **151** (1958) 174
[SF58b] SCHIRMER, H., and J. FRIEDRICH, Z. Phys. **151** (1958) 375
[SH52] SPITZER, L., and R. HÄRM, Phys. Rev. **89** (1952) 977

[Sh70] SHCHERBAKOV, A. A. (Щербаков, А. А.), J. priklad. spektr. **12** (1970) 984
[SH72a] STEPHENS, R. L., and W. F. HUG, Laser Focus **6** (1972) 38
[SH72b] SCHNEIDER, H., and E. HUGENTOBLER, Helvetica Physica Acta **45** (1972) 611
[SHB71] SCHREIBER, P. W., A. M. HUNTER and K. R. BENDETTO, Phys. Fluids **14** (1971) 2696
[SJB66] SHKAROFSKY, I. P., T. W. JOHNSTON and M. P. BACHYNSKI, The Particle Kinetics of Plasma. Addison-Wesley, Reading, Mass. 1966
[Sp56] SPITZER, L., Physics of Fully Ionized Plasmas. Interscience, New York 1956
[SP62] SCHMITZ, G., and H. J. PATT, Z. Phys. **167** (1962) 163
[SRG77] SKOWRONEK, M., P. ROMEAS and V. T. GIA, Contr. Papers, XIII ICPIG, Berlin 1977, 595
[SRGC70] SKOWRONEK, M., J. ROUS, A. GOLDSTEIN and F. CABANNES, Phys. Fluids **13** (1970) 378
[SS73] SCHIRMER, H., and J. STOBER, Z. Naturforsch. **28** (1973) 1454
[St69] STEINBERGER, S., Z. Phys. **223** (1969) 1
[Sub76] SUBAREV, D. N., Statistische Thermodynamik des Nichtgleichgewichts. Akademie-Verlag, Berlin 1976
[Suc64] SUCHY, K., Beitr. Plasmaphys. **4** (1964) 71
[Suc69] SUCHY, K., Phys. Fluids **12** (1969) 247
[Uh83] UHLENBUSCH, J., Inv. Papers, XVI ICPIG, Düsseldorf 1983, 119
[VGP71] VASILEVA, A. N., I. A. GRUGINA and V. D. PISMENNY (Васильева, А. Н., И. А. Гругина и В. Д. Письменный), Teplofiz. vysok. temp. **9** (1971) 865
[Vi71] VIEGAS, J. R., Phys. Fluids **14** (1971) 541
[VK76] VOROBEV, V. S., and A. L. KHOMKIN (Воробьев, В. С., и А. Л. Хомкин), Teor. mat. fiz. **26** (1970) 364
[VK77a] VOROBEV, V. S., and A. L. KHOMKIN (Воробьев, В. С., и А. Л. Хомкин), Teplofiz. vysok. temp. **15** (1977) 1304
[VK77b] VOROBEV, V. S., and A. L. KHOMKIN (Воробьев, В. С., и А. Л. Хомкин), Fizika plazmy **3** (1977) 885
[VK80] VECHININ, S. P., nd YU. K. KAURILENKOV (Веченин, С. П., и Ю. К. Куриленков), Teplofiz. vysok. temp. **18** (1980) 225
[VK83] VALUEV, A. A., and JU. K. KURILENKOV (Валуев, А. А., и Ю. К. Куриленков), Teplofiz. vysok. temp. **21** (1983) 591
[VKKK74] VUL, B. M., V. R. KARASIK, B. D. KONYLOVSKI, G. B. KURGANOV, V. S. VYSOTSKI, D. V. PRONKIN, YU. A. EFIMOV and G. I. AGANOV (Вул, Б. М., В. Р. Карасик, Б. Д. Коныловский, Г. Б. Курганов, В. С. Высотский, Д. В. Пронкин, Ю. А. Ефимов и Г. И. Аганов), Kvant. elektronika **9** (1974) 1983
[VMNS70] VEKHOV, A. A., G. V. MICHAILOV, F. A. NIKOLAEV and YU. P. SVIRIDENKO (Вехов, А. А., Г. В. Михайлов, Ф. А. Николаев и Ю. П. Свириденко), J. prikl. spektr. **12** (1970) 979
[VN77] VALUEV, A. A. and G. E. NORMAN (Валуев, А. А., и Г. Э. Норман), Teplofiz. vysok. temp. **15** (1977) 689
[VP72] VUKOVIĆ, S. M. and M. M. POPOVIĆ (Вукович, С. М., и М. М. Попович) Teplofiz. vysok. temp. **10** (1972) 419
[VPP72] VUKOVIĆ, S. M., S. S. POPOVIĆ and M. M. POPOVIĆ (Вукович, С. М., С. С. Попович и М. М. Попович), Teplofiz. vysok. temp. **10** (1972) 419
[VSBP79] VITEL, Y., M. SKOWRONEK, K. BENISTY and M. M. POPOVIĆ, J. Phys. **D 12** (1979) 1125
[Wi70] WILSON, J. R., J. Phys. **D 3** (1970) 2005

[Wil81a] Wilhelm, H. E., IEEE Transact. on Plasma Sci. **PS-9** (1981) 68

[Wil81b] Wilhelm, H. E., AIAA Journal **19** (1981) 533

[Wil82] Wilhelm, H. E., Plasma Physics **24** (1982) 1091

[WKH75] Wiese, W. L., D. E. Kelleher and V. Helbig, Phys. Rev. **A 11** (1975) 1854

[WKP72] Wiese, W. L., D. E. Kelleher and D. R. Paquette, Phys. Rev. **A 6** (1972) 1132

[WS74] Weisheit, J. C. and B. W. Shore, Astrophys. J. **194** (1974) 519

[WT76] Weiss, C. O. and K. Teuber, Opt. Commun. **16** (1976) 339

[Wu58] Wulff, H., Z. Phys. **150** (1958) 614

[WW69] Williams, R. H. and H. E. DeWitt, Phys. Fluids **12** (1969) 2326

[ZKN76] Zelener, B. V., Yu. K. Kurilenkov and G. E. Norman (Зеленер, Б. В., Ю. К. Куриленков и Г. Э. Норман), Khim. vysok. energi **10** (1976) 489

[ZNF75a] Zelener, B. V., G. E. Norman and V. S. Filinov (Зеленер, Б. В., Г. Э. Норман и В. С. Филинов), Teplofiz. vysok. temp. **13** (1975) 712

[ZNF75b] Zelener, B. V., G. E. Norman and V. S. Filinov (Зеленер, Б. В., Г. Э. Норман и В. С. Филинов), Teplofiz. vysok. temp. **13** (1975) 913

9. Subject Index

Bei Fragen zur Produktsicherheit wenden Sie sich bitte an:
If you have any questions regarding product safety,
please contact:

Walter de Gruyter GmbH
Genthiner Straße 13
10785 Berlin
productsafety@degruyterbrill.com